THE FIRST SHOTS

*The Epic Rivalries
and Heroic Science
Behind the Race
to the
Coronavirus
Vaccine*

THE
FIRST
SHOTS

|||||||||||||||||||||||||||||||||||

Brendan Borrell

SUGAR 23 BOOKS   **MARINER BOOKS**
An Imprint of HarperCollins*Publishers*
Boston   New York

marinerbooks.com

Library of Congress Cataloging-in-Publication Data has been applied for.
ISBN 978-0-358-56984-8 (hardcover)
ISBN 978-0-358-56618-2 (ebook)

Book design by Chrissy Kurpeski

Printed in the United States of America
1 2021
4500837338

It is not the critic who counts; not the man who points out how the strong man stumbles or where the doer of deeds could have done them better. The credit belongs to the man who is actually in the arena, whose face is marred by dust and sweat and blood; who strives valiantly; who errs, who comes short again and again . . . who knows the great enthusiasms, the great devotions . . . who at the best knows in the end the triumph of high achievement, and who at the worst, if he fails, at least fails while daring greatly, so that his place shall never be with those cold and timid souls who neither know victory nor defeat.

— THEODORE ROOSEVELT, APRIL 23, 1910,
A version of which is on the wall of Barney Graham's office at the
Vaccine Research Center of the National Institutes of Health

CONTENTS

PART 3: TESTING

May 2020–October 2020

PART 4: ROLLOUT

October 2020–January 2021

CAST OF CHARACTERS

During the Coronavirus Pandemic Year of 2020

White House

Donald Trump President
Michael Pence Vice president
Mark Meadows Chief of staff to the president*
Joseph Grogan Director, Domestic Policy Council
Deborah Birx Coordinator, White House Coronavirus Task Force
Jared Kushner Senior adviser to the president; Trump's son-in-law
Adam Boehler CEO, U.S. International Development Finance
Corporation; Kushner's friend

U.S. Department of Health and Human Services (HHS)

HEADQUARTERS, WASHINGTON, DC
Alex Azar Secretary
Paul Mango Deputy chief of staff for policy
Brett Giroir Assistant secretary for health
Michael Caputo Assistant secretary for public affairs

---

* Mark Meadows replaced Mick Mulvaney as chief of staff on March 31, 2020.

Robert Kadlec Assistant secretary, Office of the Assistant Secretary for Preparedness and Response (ASPR)

Michael Callahan Special adviser to the ASPR; staff physician, Massachusetts General Hospital

Rick Bright Director, Biomedical Advanced Research and Development Authority (BARDA)

NATIONAL INSTITUTES OF HEALTH (NIH), BETHESDA, MARYLAND

Francis Collins Director

Anthony Fauci Director, National Institute of Allergy and Infectious Diseases (NIAID)

John Mascola Director, Vaccine Research Center, NIAID

Barney Graham Deputy director, Vaccine Research Center, NIAID

Kizzmekia Corbett Research fellow, Vaccine Research Center, NIAID

Jason McLellan Former research fellow; current associate professor, University of Texas at Austin

Larry Corey Co-leader of the NIAID-funded COVID-19 Prevention Network; former president, Fred Hutchinson Cancer Research Center

CENTERS FOR DISEASE CONTROL AND PREVENTION (CDC), ATLANTA, GEORGIA

Robert Redfield Director

Anne Schuchat Principal deputy director

Nancy Messonnier Director, National Center for Immunization and Respiratory Diseases

FOOD AND DRUG ADMINISTRATION (FDA), SILVER SPRING, MARYLAND

Stephen Hahn Commissioner

Peter Marks Director, Center for Biologics Evaluation and Research (CBER)

OPERATION WARP SPEED, WASHINGTON, DC

General Gustave Perna Chief operating officer

Moncef Slaoui Chief scientific adviser

The Wolverines

James Lawler Director, Clinical and Biodefense Research, University of Nebraska Medical Center

Matthew Hepburn Vaccine lead, Operation Warp Speed

Richard Hatchett CEO, Coalition for Epidemic Preparedness Innovations

Carter Mecher Senior medical adviser, Department of Veterans Affairs

The Vaccine Makers

MODERNA THERAPEUTICS
mRNA Vaccine

Stéphane Bancel CEO

Stephen Hoge President

Tal Zaks Chief medical officer

Derrick Rossi Cofounder

PFIZER-BIONTECH
mRNA Vaccine

Albert Bourla CEO, Pfizer

Mikael Dolsten Chief medical officer, Pfizer

Katalin Karikó Senior vice president for therapeutics, BioNTech

NOVAVAX
Protein-Subunit Vaccine

Stanley Erck President and CEO

Gregory Glenn Chief scientific officer

ASTRAZENECA
Viral Vector Vaccine

Menelas Pangalos Executive vice president, biopharmaceuticals R and D

Sarah Gilbert Professor, Jenner Institute, University of Oxford

JOHNSON AND JOHNSON
Viral Vector Vaccine
Alex Gorsky CEO

SANOFI
Protein-Subunit Vaccine
Paul Hudson CEO

MERCK
Live Attenuated Chimeric Vaccine
Ken Frazier CEO

PROLOGUE

上海市公共卫生临床中心
SHANGHAI PUBLIC HEALTH CLINICAL CENTER
PEOPLE'S REPUBLIC OF CHINA
1:30 P.M., FRIDAY, JANUARY 3, 2020

The specimen vial came sealed in a metal box. It had been sent here from the city of Wuhan, some five hundred miles inland, under desperate circumstances. On December 26, 2019, a man had shown up at Wuhan's Central Hospital with a cough and a high fever. The patient was forty-one years old and worked at the Huanan Seafood Market, a bustling complex with over a thousand stalls. There, beneath the skyscrapers of that city of eleven million, were stacks of cages trucked in from the countryside containing animals like bamboo rats and raccoon dogs, an Asian fox relative with spectacled eyes. Some dealers even offered cobras, which are prized as both food and medicine. These captives were all waiting to be butchered and then thrown over a wood fire or boiled in a salty broth.

Winter in Wuhan meant foggy days, light rain, and temperatures in the forties—cold season. The market worker had been feeling ill for nearly a week. His cough progressed to a profound pain in his chest. Tests for bacterial pneumonia and flu came back negative. So, too, did tests for other common respiratory pathogens. After three days, he was rushed to the hospital's intensive care unit, gasping for air. Doctors put him under anesthesia and inserted the long, clear tube of a mechanical ventilator down his trachea. The breathing machine would perform the work that his lungs no longer could. A second tube snaked through his nose and into his stomach to provide sustenance. There was, for a short time, a third tube, this one fed down the trachea and into one of the most distant branches of his lungs. Through this tube, doctors squirted sterile saline and suctioned a foamy, pinkish fluid back into a vial. The lung-wash sample was preserved in alcohol, packed in the metal box with dry ice, and sent to the public health center on the outskirts of Shanghai where a microbiologist named Yong-Zhen Zhang was waiting.

As a child, Zhang had wanted to be a general leading soldiers into battle. Instead, he became a scientist committed to protecting the health of his countrymen and demonstrating China's scientific competency on the world stage. Back when he worked for the Chinese Center for Disease Control and Prevention in Beijing, it wasn't unusual for colleagues to get e-mails from him at three a.m. or for visitors to come by his office for morning tea and find him curled up on a small black leather couch where he had slept the night before.

Zhang knew that mysterious cases of acute pneumonia had been trickling into Wuhan's hospitals since the beginning of December, perhaps even earlier, but local officials had not been forthright about what was happening. They initially decided against notifying federal authorities and ordered samples of the virus destroyed. Then, on December 30, a notice appeared on the website of the Wuhan city government:

市卫生健康委关于报送不明原因肺炎救治情
　　况的紧急通知

各有关医疗机构:
　　根据上级紧急通知,我市华南海鲜市场陆续出现不明原 因肺炎病人。为做好应对工作,请各单位立即清查统计近一 周接诊过的具有类似特点的不明原因肺炎病人,于今日下午 4点前将统计表(盖章扫面件)报送至市卫健委医政医管处 邮箱。
　　联系人: 毛冰 85697893 李莹 85690943
　　邮箱: whsyzc@126.com

附件: 相关信息上报表

武汉市卫生健康委医政医管处

It was an "emergency notification" from the city's health commission addressed to "all relevant medical institutions" about "a pneumonia of an unknown cause." The health commission was asking hospitals to submit statistics on any patients whose illnesses fit that description by four o'clock that afternoon. A second notice told health workers to prepare for a potential outbreak by organizing expert teams, rapidly diagnosing patients, and devising a cohesive treatment and infection-control plan. "No unit or individual without authorization may release medical treatment information," read the mandate.

The next day, local officials released more information, and a five-sentence story headlined "China Investigates Respiratory Illness Outbreak Sickening 27" went out on the international newswire. Countries that are members of the World Health Organization—a Geneva-based agency of the United Nations—are obligated to share details about disease outbreaks. When the WHO asked China about the report, government officials said not to worry. Like rabies or West Nile virus, this new pathogen seemed to have leaped from animal to human, but there was no evidence that it was now spreading from person to person. Authorities shut the market on January 1 and told the WHO that none of their health-care workers had been infected.

Two days later, in Shanghai, inside a building with a red pagoda-style roof, Zhang and his team opened up the metal box and got to work inside their institute's airtight biosafety level 3 laboratory. The BSL-3 is the second-most-secure laboratory type, reserved for potentially lethal viruses that may spread in the air. Entering it required passing through two self-closing doors. Once inside, technicians conducted their work with their gloved arms inside a glass cabinet that sucked out and filtered any contaminated air.

The sample in Zhang's hands was teeming with bacteria and viruses. Over the weekend, Zhang separated out the great abundance of human genes in the lung wash and then sequenced the remaining gene fragments. Many were from organisms that live in the human mouth, in-

cluding the bacteria that form dental plaques and those that cause gin-
givitis. Zhang also saw, in high numbers, genes from an opportunistic
bacteria that might occasionally cause pneumonia, but only in someone
who was already sick. Buried in all of these generally harmless passengers in the human body was one long sequence of RNA that there was
no mistaking for anything else: a coronavirus.

Zhang had, in recent years, begun to specialize in RNA viruses, a
murderers' row of some of humanity's worst foes. These range from
the common cold and influenza viruses to Ebola, human immunode-
ficiency virus (HIV), and the coronavirus that caused the outbreak of
severe acute respiratory syndrome (SARS) in 2002. These viruses carry
their payload on a single strand of ribonucleic acid (RNA), an information molecule that has a number of uses in cells, including as a template
for producing proteins.* Although viruses have the same universal genetic code that we do and are made up of the same chemical elements
that we are, many scientists would argue that a virus is not a living thing.
Unlike bacteria or the single-celled parasites that cause malaria, viruses
have no ability to move, metabolize, or replicate on their own. Viruses
are too small to see under a microscope. The largest ones are the size of
the smallest bacteria, but most are a hundred times smaller. They depend entirely on their hosts' cells following their instructions. Inside a
cell, a single virus can make thousands of copies of itself, which bud out
of the cell membrane one by one or burst free all at once upon the death
of that cell. Then they fan out to do it all over again. Within days, the
number of viruses in the host can rise to the trillions. As the British biologist Peter Medawar is said to have quipped, a virus is "a piece of bad
news wrapped in a protein."

The first human coronaviruses were discovered in the 1960s, and
a total of four of them were later found to commonly infect people,
though they rarely produced more than the sniffles. Then, in November
2002, SARS emerged in southern China. Symptoms included a fever and
a cough that could progress to a full-blown lung infection and, potentially, heart and liver failure. Over the next eight months, 8,437 people

---

* Other viruses carry their genetic material on DNA or even double-stranded RNA, a
biological rarity.

developed the illness and 813 died—a frightening and chaotic episode scorched into Zhang's memory. When a coronavirus was identified as the culprit, it was a wake-up call that these pathogens were a threat requiring the same diligence the Chinese had given the U.S. submarines lurking beneath the South China Sea. The SARS virus was traced to Himalayan palm civets, mongoose-like mammals that were sold in local markets around Hong Kong. Those civets appeared to have been infected by wild bats.

Zhang could see that this new coronavirus from Wuhan shared much of its genome with a coronavirus that had been isolated from a bat in the subtropical province of Yunnan, near the Vietnam border. The two diverged over time and distance with each new victim they infected. Zhang completed his analysis on January 5 and hoped to publish the new coronavirus gene sequence as soon as possible. He was one of several Chinese experts who had obtained the genome by then, but the news was kept under wraps. That week, Wuhan police were rounding up local doctors who had posted about the SARS-like coronavirus on social media. The authorities admonished them for "rumormongering" on the internet—a crime in China. The Ministry of Health issued a gag order on publishing the genome, but the drip-drip-drip of information about the outbreak continued online. When the Chinese government belatedly confirmed the recent illnesses were linked to a coronavirus, the WHO was effusive with its praise, stating that the sequencing was "a notable achievement and demonstrates China's increased capacity to manage new outbreaks."

That sequence, however, was still not online. "It just seemed ridiculous," said Zhang's collaborator and close friend Eddie Holmes, who was based in Melbourne, Australia. On the morning of January 11, Zhang was sitting on the runway at the Shanghai airport preparing to fly to Beijing on a work trip when Holmes rang his cell phone. Holmes had been privy to the analysis but not the actual sequence. He told Zhang that researchers around the world were clamoring for it, and he should release it. "I asked Holmes to give me one minute to think," Zhang said. A vertical crease between Zhang's eyebrows deepened whenever he was deliberating over something, as he was then. Holmes waited. The engines were revving up.

"We need to release this now," Holmes insisted.

"Okay, okay, okay," Zhang said. He hung up and e-mailed Holmes the file, called WH-Human_1.

Holmes then e-mailed a friend in Edinburgh, Scotland, Andrew Rambaut, who moderated a website called Virological.org, a virtual watering hole for virus nerds, and told him he had the genome sequence. It was already after midnight in Edinburgh, but Rambaut was a night owl. He told Holmes to send the file his way and they hashed out a short description of it, being careful to avoid violating the privacy of the patient, who had recovered. Finally, Rambaut hit Upload. It was a Friday evening in the United States, but links to the sequence soon ricocheted on Twitter and were shared among hundreds of researchers in e-mail chains, text messages, and Slack groups.

"And so it begins," wrote one virologist.

"Traveling at the speed of science," wrote another.

"Amazing times," wrote a third.

There would be no celebrating for Zhang, however. The Chinese patriot would pay dearly for violating the government's gag order as authorities swooped in and—one must infer because Zhang has never said it directly—confiscated the remnants of the lung-wash sample in his freezer. They clamped down on his lab's activities and treated him as an enemy of the state. *Rectification* is the official word for it. "The shit hit the fan" is how Holmes put it. As China cracked down inside its borders, it told the world that it hadn't seen a new case in over a week. Since the outbreak began, officials said, 41 people had been preliminarily diagnosed and isolated, and 763 of their contacts had been traced. The coronavirus, in other words, was contained.

PART 1

EMERGENCE

January 2020–March 2020

1

SOUNDS LIKE IT
COULD BE FUN

Four days before Zhang released the sequence of the new coronavirus, Jason McLellan was standing inside a crowded ski shop in
Park City, Utah, waiting to get a pair of new snowboarding boots
heat-molded to his feet. His phone buzzed in his pocket. He glanced
down at the screen. It was his former mentor Barney Graham, a vaccine researcher at the National Institutes of Health. McLellan had left the
NIH seven years earlier and was now an associate professor in Austin,
Texas, albeit the kind of professor who gives lectures in a pair of shorts
and a *Star Wars* T-shirt.

McLellan slipped outside to take the call. The pneumonia cluster in
Wuhan was potentially caused by a coronavirus, Graham said. He was
making plans to design and test a vaccine for it as quickly as possible,
and he wanted McLellan to sign on. At that particular moment, neither
thought that humanity was necessarily facing a pandemic. It was in the
back of Graham's mind, for sure, but it was always in the back of Graham's mind, considering his line of work.

A doctor and virologist, Graham was an unassuming Midwesterner,
six foot five with broad, sloping shoulders and a gray beard. For nearly
twenty years, Graham had worked at the NIH's Vaccine Research Center (VRC), where he studied the way certain viruses caused the immune
system to go haywire. He got a far-off look in his eyes when he spoke
about the viruses he had tangled with, evincing the grudging respect
of a rancher for the coyote that has taken another calf. When he turned

sixty, the onetime farm boy from Kansas bought himself an F-150 pickup truck with burgundy seats. His license plate: VAXIN8R.

Located nine miles from the U.S. Capitol, the VRC—in Building 40 on the NIH's grassy, university-like campus in Bethesda, Maryland—was originally established to address the HIV/AIDS pandemic, but Graham delved into lesser-known diseases. He knew that the world had seen viruses spill over from wild-animal populations into humans many times, with devastating consequences. HIV had come from chimpanzees, likely via a hunter who had infected himself while butchering an animal for its meat.

As Graham had discussed with Anthony Fauci, the renowned infectious diseases doctor who had recruited him to the NIH, there were certain things that scientists could do to help prepare humanity for the next outbreak. Animal-borne diseases that once might have been confined to a remote cave in the middle of a tropical forest now posed global threats, given that it was easier than ever for a person just about anywhere to hop on a bus in the morning and be on an international flight by the afternoon. In the past decade, the WHO had declared four public health emergencies related to zoonotic diseases: swine flu in 2009, Ebola in 2014 and 2018, and Zika in 2016. What would the next one be?

Some experts referred to the potential danger on the horizon as disease X, and Graham had an idea about how we—humanity—might work together to stop it. There were about 120 viruses known to pose a potential risk to people. Those viruses could be divided into twenty-five different families. There were licensed vaccines for viruses in just thirteen of those families. Most global-health efforts focused on the troubling pathogens that had emerged in the past—but that left a whole swath of the viral landscape unknown, unprobed, unprotected against. Over the past decade, Graham had been advocating for also studying representatives in the twelve virus families for which there were no vaccines. By developing a vaccine strategy against these "prototype pathogens," researchers would have a head start if any of their close relatives emerged one day. Instead of being one step behind an outbreak, Graham wanted humans one step ahead. Going after this new coronavirus in China was going to be a flex, a demonstration of how fast science could move in the face of a new threat.

Graham knew there was no one faster than Jason McLellan. McLellan was Graham's structural-biology guru, the person who could tell him the precise shape of the different parts of this virus.

After McLellan hung up the phone, he fired off a WhatsApp message to one of his brightest graduate students back at the University of Texas at Austin, Daniel Wrapp, who had been involved in a previous coronavirus vaccine study a few years earlier. "Graham is going to try and get the coronavirus sequence out of Wuhan China," he wrote. "He wants to rush a structure and vaccine, etc. You game?"

Wrapp responded a minute later. "Should be straightforward other than the rush," he wrote. "Sounds like it could be fun."

Back at the NIH, Graham thought he had an inside line on the coronavirus sequence, which he would need for the vaccine work to begin. At Fauci's suggestion, he reached out to George Gao, the head of China's CDC, a scientist whom Graham had met at several international meetings. Days went by without a response. In the end, the NIH obtained the sequence when it went global on Virological.org on January 10. "It's online so everyone has it now," McLellan griped in a message to a member of his lab in Texas.

At around eight o'clock on Saturday morning, Wrapp rolled out of bed at his home in Austin's West Campus neighborhood and made a beeline for the laboratory. He plopped his MacBook on his desk next to the window at the end of a long black lab bench scattered with chemical reagents and pipettes. He put on a pot of coffee, poured himself a cup, and began to work. The entire program for the novel coronavirus came down to a string of letters, nearly thirty thousand of them in all. *Bad news wrapped in a protein.* The string began like this:

```
ATTAAAGGTTTATACCTTCCCAGGTAACAAACCAA
CAACTTTCGATCTCTTGTAGATCTGTTCTCTAAAC
GAACTTTAAAATCTGTGTGGCTGTCACTCGGCTGCATGCT
TAGTGCACTCACGCAGTATAATTAATAACTAATTACTGTC
GTTGACAGGACACGAGTAACTCGTCTATCTTCTGCAGGCT
GCTTACG . . .
```

Each letter in that sequence represented one of four nucleotide bases, the molecules that make up the universal genetic code: A for adenine,

G for guanine, C for cytosine, and then T, which, in this particular case, represented uracil.* In the coronavirus, those instructions are strung together as a ribbon of ribonucleic acid, or RNA. The RNA molecule is similar to its double-stranded cousin deoxyribonucleic acid, or DNA, but RNA consists of only a single chain of nucleotides. A cell reads the RNA as a series of three-letter words, known as codons. There are exactly sixty-four codons — every possible three-letter combination of the four bases. The codon ATG signifies the beginning of a gene. Three other codons indicate the end. The remaining sixty codons correspond to the twenty amino acids that cells can produce, with some redundancy baked into the system.

It is these amino acids that serve as the building blocks for proteins, molecules that you probably think of as just another part of your diet, measured by the gram on the nutritional label of your yogurt container. When you digest proteins, your body breaks them down into individual amino acids. Then your genes tell your cells how to string together new proteins, a nearly endless variety of clamps and springs and zippers that fill the toolbox used by all living and not-quite-living things. Antibodies, for instance, are proteins that the body's immune system produces to glom on to viruses and disable them. So, too, are most of the bits and pieces of viruses themselves. In a virus, some proteins are structural, forming a capsule. Others are enzymes, chemicals that help the virus make copies of itself.

A gene is a specific sequence of RNA or DNA that codes for a protein. This new virus had exactly sixteen genes, and Wrapp had been sent the sequence of just one of those in an e-mail. It began at the 21,563rd base. This gene coded for the forty or so spikes that stud the surface of the virus like a crown, giving this family of pathogens its name: coronavirus. These spikes are less a regal adornment than a lockpick, perhaps even a switchblade that mostly stays folded to avoid detection by the immune system. Each spike is made up of three identical proteins braided together. When the tip of a coronavirus spike hinges open, the

---

* Thymine is found only in DNA, while uracil occurs only in RNA. Both nucleotide bases are represented as a T in data files and software programs.

virus can dock on a human cell. The exterior of every human cell is covered in various types of receptors, some of which act like keyholes. Different viruses have evolved to pick the locks of different keyholes. After the coronavirus spike protein binds to its chosen receptor, it thrusts outward, fuses to the cell membrane, and injects its nefarious payload. The first SARS virus bound to the ACE2 receptor, common on the surface of cells in the lungs. Once inside, the viruses take over the lungs and other parts of the body, eating away at the host and turning it into them.

Wrapp opened the spike sequence using a program called Geneious. The sequence began with an ATG, which coded for the amino acid methionine. The next three-letter codon was TTT, which was phenylalanine. Then came GTT, for valine. It went on like this for 3,819 letters until it reached TAA, the stop codon. String those 1,273 amino acids together and you have it, the entire spike protein:

MFVFLVLLPLVSSQCVNLTTRTQLPPAYTNSFTRGVYYPD
KVFRSSVLHSTQDLFLPFFSNVTWFHAIHVSGTNGTKRFD
NPVLPFNDGVYFASTEKSNIIRGWIFGTTLDSKTQSLLIVN
NATNVVIKVCEFQFCNDPFLGVYYHKNNKSWMESEFRVYS
SANNCTFEYVSQPFLMDLEGKQGNFKNLREFVFKNIDGYF
KIYSKHTPINLVRDLPQGFSALEPLVDLPIGINITRFQTLLAL
HRSYLTPGDSSSGWTAGAAAYYVGYLQPRTFLLKYNENGTIT
DAVDCALDPLSETKCTLKSFTVEKGIYQTSNFRVQPTESIVR
FPNITNLCPFGEVFNATRFASVYAWNRKRISNCVADYSV
LYNSASFSTFKCYGVSPTKLNDLCFTNVYADSFVIRGDEVRQ
IAPGQTGKIADYNYKLPDDFTGCVIAWNSNNLDSKVG
GNYNYLYRLFRKSNLKPFERDISTEIYQAGSTPCNGVEGFN
CYFPLQSYGFQPTNGVGYQPYRVVVLSFELLHAPATVCGP
KKSTNLVKNKCVNFNFNGLTGTGVLTESNKKFLPFQQFGRDI
ADTTDAVRDPQTLEILDITPCSFGGVSVITPGTNTSNQVAV
LYQDVNCTEVPVAIHADQLTPTWRVYSTGSNVFQTRAG
CLIGAEHVNNSYECDIPIGAGICASYQTQTNSPRRARSVASQ
SIIAYTMSLGAENSVAYSNNSIAIPTNFTISVTTEILPVSMT
KTSVDCTMYICGDSTECSNLLLQYGSFCTQLNRALTGIA
VEQDKNTQEVFAQVKQIYKTPPIKDFGGFNFSQILPDPSKP

SKRSFIEDLLFNKVTLADAGFIKQYGDCLGDIAARDLICAQK
FNGLTVLPPLLTDEMIAQYTSALLAGTITSGWTFGAGA
ALQIPFAMQMAYRFNGIGVTQNVLYENQKLIANQFNSAIG
KIQDSLSSTASALGKLQDVVNQNAQALNTLVKQLSSN
FGAISSVLNDILSRLDKVEAEVQIDRLITGRLQSLQTYVTQQLI
RAAEIRASANLAATKMSECVLGQSKRVDFCGKGYHLMS
FPQSAPHGVVFLHVTYVPAQEKNFTTAPAICHDGKAHF
PREGVFVSNGTHWFVTQRNFYEPQIITTDNTFVSGNCDV
VIGIVNNTVYDPLQPELDSFKEELDKYFKNHTSPDVDLGDIS
GINASVVNIQKEIDRLNEVAKNLNESLIDLQELGKYEQYIK
WPWYIWLGFIAGLIAIVMVTIMLCCMTSCCSCLKGCCSCG
SCCKFDEDDSEPVLKGVKLHYT

In another window, Wrapp pulled up the amino acid sequence for the original SARS virus. There was a more than 80 percent match. "Spike sequence looks really similar to SARS, to the point that it probably even binds to ACE2," he wrote McLellan.

Wrapp copied and pasted the entire spike sequence into an empty Microsoft Word document, then scanned the spike for a lysine (K) and a valine (V) together. He found it about three-quarters of the way through:

FGAISSVLNDILSRLD**KV**EAEVQIDRLITGRLQSLQTYVTQQLI

He moved his cursor over those two letters and replaced them with two Ps, for the amino acid proline. This change in the spike's makeup, Wrapp knew, would likely prevent the spike from thrusting outward as it does on a live virus during the fusion process. Frozen in this state, the modified spike would leave exposed its most vulnerable region, the target that the immune system needed to learn to attack. That is the fundamental idea of a vaccine. It is a dummy round injected into the body in some form so that the immune system will be prepared for a real invasion, be it viruses, bacteria, or other microorganisms, such as the malaria parasite.

The two-letter change that Wrapp made represented a decade's worth of discoveries from the Graham and McLellan labs compressed into the span of ten minutes. He attached the sequence for the stabilized spike, termed the S-2P design, to an e-mail and sent it to the NIH at 9:26 a.m. Then he poured himself a second cup of coffee.

Graham spent the weekend mulling over other features of the vaccine design before settling on a final sequence. On Monday, January 13, he walked across the NIH campus to Building 31, the executive offices. He was meeting with Tony Fauci. The son of a pharmacist in Brooklyn, Fauci had made his name studying the tricks HIV uses to break down the body's defenses. He was seventy-nine, but he was as sharp as ever, a walking multiprocessor with darting eyes and the decisiveness of an army general. This was a guy who had been the captain of his high-school basketball team even though he was one of the shortest players on the court. He had been offered the directorship of the NIH multiple times, a political appointment, but he had chosen to remain the head of one of its institutes, the National Institute of Allergy and Infectious Diseases, or NIAID, for more than thirty years. It was a post that gave Fauci the ability to speak his mind without fear of repercussions. In a visit to the Clinton White House in December 1996, Fauci had proposed creating a research center to develop vaccines against AIDS. A quarter of a century later, Graham told Fauci the Vaccine Research Center was wielding all the scientific tools it had developed for that unfinished mission against this new foe.

But science alone was not going to be enough. The NIH team had to select the right technology to deliver its spike-based vaccine to the body. Graham did not need to remind Fauci that he and the director of the VRC, John Mascola, had struggled to find a commercial manufacturing partner during the Zika outbreak several years earlier. But at the last minute, the VRC scientists were asked to run tests on monkeys with a vaccine produced by a biotech company called Moderna and were galvanized by those results. In the fall of 2019, the company's CEO, Stéphane Bancel, had offered to manufacture a pilot lot for the next disease Graham wanted to attack for his pandemic demonstration project. On January 7, Bancel, at his vacation home in the South of France, read the news about the coronavirus. He had e-mailed Graham, eager to begin.

Fauci had only one question for Graham: "Do you need money for the pilot lot?"

"No," Graham said. The VRC had already set aside some money for this program, and Bancel was looking for more. He was always looking for more.

The scientific kickoff meeting at the NIH took place the following day, January 14, at 11:00 a.m. One of Graham's postdoctoral researchers, Kizzmekia Corbett, had rounded up as many academic collaborators as she could, and she asked Graham to nudge any professors who'd failed to respond to the Zoom invite. Corbett was a prolific Twitter user; her profile @KizzyPhD featured a picture of her in a black dress, a faux-pearl necklace, and fiery red lipstick, and her bio read "Virology. Vaccinology. Vagina-ology. Vino-ology. My tweets are my own. My science is the world's." Corbett's inch-long pink fingernails were hard to miss, but it was the color of her skin that stood out most. She was the rare Black woman in a field dominated by white and Asian dudes, and she wasn't afraid to own it.

Corbett had grown up at the end of a dirt road in Hurdle Mills, North Carolina, where her elementary school was surrounded by tobacco farms. A gifted science student in high school, she obtained a prestigious scholarship for minorities that brought her to the University of Maryland. During college, she spent her summers working at the NIH, which is how she first got to know Graham. He once asked her where she wanted to be in ten years. "I want your job," she told him. Graham knew he would have to hire her one day. "She was really pretty bold," he said. She later earned her doctorate in the laboratory of virologist Ralph Baric at the University of North Carolina. "The O.G. in the coronavirus field," Corbett called him.

During the kickoff meeting, Corbett told the team that Moderna had received the S-2P sequence on Monday and would begin to manufacture its first doses of vaccine. The company could be shipping the pilot lot to the NIH's clinical trial site within four months — by early May. The collaborators on the call were tasked with running the gauntlet of preclinical tests needed before a single dose was injected during the first small-scale human-safety study, known as a phase 1 trial. Their animal studies would tell the researchers whether the vaccine was likely to work and, later, whether it was safe enough to expand to larger clinical trials.

The NIH collaborators would also be developing and validating the assays needed for the human study. None of this was new for them, but the way they were doing it was: completely in parallel. It was as if they were building all the parts of a new car simultaneously, with one team working on the engine, another on the suspension, and someone else planning a road trip before any of that was finished. That was what it was going to take to fight disease X.

The critical first step for everyone was growing gobs of the spike protein in specialized cell cultures. Jason McLellan's lab in Texas would then obtain images of the spike to see if the sequence tweaks had worked to immobilize it as predicted. Corbett and Graham, meanwhile, would be testing mice to see whether and to what extent the vaccine stimulated the production of antibodies against the spike protein. John Mascola's lab would be running the same antibody assays when samples came in from the human study. Next, Mark Denison, Graham's buddy from medical school, and his colleagues at Vanderbilt University in Nashville would test whether those antibodies, isolated from human blood serum, could neutralize the live coronavirus with its functional spikes. Finally, there was Corbett's mentor Ralph Baric. Researchers in Baric's lab would be doing the dangerous work inside a biosafety level 3 lab: challenging vaccinated mice with live coronavirus and seeing how well they fended it off.

When the meeting was over, all the members of the team knew what they had to do. They were still just a bunch of preppers running a fire drill, making sure all their procedures were in place in case this outbreak didn't burn out on its own. There had been a few short news segments on the coronavirus, but most of the world was still not paying attention to the virus or what scientists were up to, which was how it usually went with science.

In order to make the spike protein, the digital gene sequence that Wrapp had created had to be "printed out" in the form of DNA. Almost everyone in the business outsourced this process, typically to GenScript, a company that made ordering a piece of DNA almost as easy as ordering groceries on Instacart. It was another sign of the interconnectedness of the world that the company so many U.S. researchers relied on was based in China, though it did have operations in Piscataway, New Jersey.

Corbett had submitted the team's spike-sequence orders a few days before the NIH conference call, on Saturday, January 11. Over the weekend, a researcher on the Texas team, Nianshuang Wang, became concerned. "You synthesized the whole gene in GenScript?" he wrote in a WhatsApp message to McLellan. He feared that GenScript was going to be hit with a crush of orders because so many scientists would be seeking supplies for their coronavirus work. Even though the company had had its Chinese employees cancel their vacations, orders were undoubtedly going to pile up. Getting the spike-protein DNA, Wang warned, "may take several weeks."

Wang had an idea. Rather than ordering one big loop of DNA — known as a plasmid — that would take a while to arrive, he could order five or six smaller gene fragments that would be shipped out in two to four days. Wang would have to stitch the whole thing together in the Austin lab, a laborious, technically complex process but one that might save the team some time. Wang, who had been raised by peasant farmers in a mud-brick home in eastern China, was undeterred by a little hardship. "We should try all efforts," he wrote.

Wang worked day and night, nodding off on a rolling office chair between steps in the process. He was making nine different versions of the spike — or fragments of the spike — for his team's different assays. In nine days, he had the complete constructs in hand, beating the arrival of the first GenScript order Corbett had submitted by a full two weeks. It gave the team a huge head start during an outbreak that was growing worse by the day. Wang sent the NIH the DNA that would allow Corbett's team to grow spike proteins and develop tests to measure the immune response to the vaccine in mice, monkeys, and, potentially, people. The urgency at that early stage came more from the scientific competition to be the first to publish details of the new coronavirus rather than from any certainty that a vaccine would be required to stop the outbreak. It was still just a fire drill, right?

2

COURAGE AND DOUBT

During the third week in January in Davos, Switzerland, when snow piles up in the streets and the World Economic Forum is in session, the regional airports become so congested with the traffic of private jets that a nearby military base opens an airstrip for them. At least 119 billionaires were on the guest list in 2020 to talk about big ideas for changing the world and to schmooze with world leaders as well as members of lowlier economic classes, including teenage climate activist Greta Thunberg and cellist Yo-Yo Ma.

Over the previous week, the number of cases of what some news anchors were calling the "Wuhan flu" jumped from 41 to 282, with a handful of infections confirmed in Thailand, Japan, and South Korea. When a reporter from the financial network CNBC brought up the topic to Pascal Soriot, the CEO of the British pharmaceutical giant AstraZeneca, he appeared unperturbed. "It really looks like at this point it's very contained," he responded. Asked about the potential impact of the coronavirus on the economy, Paul Hudson, CEO of Sanofi, one of the world's top vaccine producers, changed the subject to discuss his company's seasonal flu shots, a reliable profit-earner. "We sprint to make sure that we can get enough doses," he said. The venerable American company Merck told the network it would wait for guidance from the WHO.

Big Pharma had, by and large, concluded that there was no upside to going after emerging infectious diseases. As Merck and Sanofi had learned firsthand from their own experiences with Ebola and Zika, most outbreaks are terrifying, headline-grabbing events that quickly become utterly forgettable — at least to those not directly affected. Vaccines cost hundreds of millions of dollars to develop and often take years, decades even, to bring to market. Very few vaccines share the history of Merck's 1967 mumps vaccine, developed and approved in four years and still needed today.* Once a fleeting epidemic has passed, there is no one to test your vaccine candidate on and, more to the point, no one to sell it to. The external funding from governments and foundations, if there ever was any, dries up, and the project gets shelved. Your highly paid scientists go back to the moneymaking projects that were disrupted by the drama — and for what?

Moderna CEO Stéphane Bancel was well aware of that dismal record when he arrived in Davos that week with a single-minded goal: to secure funding for the development of his company's coronavirus vaccine. With the metal-framed glasses of an engineer and a runner's body under his well-tailored business suit, Bancel fit right in at the annual Davos meetings, which he had attended since his thirties. Now in his forties, he wasn't himself a billionaire, but he wasn't far off either. If there was one book that would stick with him that year, it was *The Ride of a Lifetime* by former Disney CEO Bob Iger, who wrote about how playing it safe never got anyone anywhere in the business world: "True innovation occurs only when people have courage."

Just as Barney Graham was out to prove that vaccine design could move faster than ever under his pathogen-preparedness model, Bancel saw an opportunity to demonstrate that his company had a technology — messenger RNA — to turn that design into reality faster than its competitors. Most vaccines developed since the 1980s depended on a fussy and time-consuming process of brewing virus proteins in large stainless-steel vats filled with a genetically engineered broth of cells. Moderna's method instead delivered to human cells only the instructions they needed to make those proteins. The vaccine, like the coronavirus itself, hijacks the machinery that already exists in our cells.

---

* Maurice Hilleman, the most prolific vaccine maker of the twentieth century, famously isolated the live mumps virus from his own daughter, Jeryl Lynn, in 1963 and weakened it by growing it in chicken eggs.

You probably know that your genome, the sum total of your genes, is contained on DNA, which takes the shape of a double helix, a twisted-rope-ladder-like structure. The second strand serves as a backup drive for the first, forming a redundant and durable archive containing all the genes you need to live. One gene, for instance, describes how to make insulin, an enzyme that controls blood sugar; another contributes to the amount of pigment in your eyes. The twenty-three pairs of chromosomes tangled inside the cell's nucleus are made of DNA, but DNA doesn't actually *do* much inside a cell. That role falls to DNA's hyperactive sidekick mRNA, which puts those instructions into action at just the right time.

When a specific gene needs to be expressed — that is, turned into protein — the two strands of DNA temporarily untwist and peel open like a zipper. Nucleotides floating in the cell like Scrabble letters then arrange themselves along these exposed strands, forming a messenger RNA that corresponds with a specific gene. These short-lived, single-stranded molecules then charge out of the nucleus and link up with ribosomes, cellular factories that look like stacks of pancakes. Inside the ribosome factories, these messenger RNAs deliver instructions for the piecing together of the needed proteins.

Moderna's vaccines hack this process, delivering synthetic mRNA straight into the cells. The ribosomes don't care where those foreign mRNA strands came from. For several days, perhaps as long as a week, until those mRNA strands decay, the ribosomes follow their instructions and output a protein. That protein serves as a dummy molecule on which the body can hone its immune response.

It was a thrilling idea, but would it work? Messenger RNA had never been approved for use in humans as a vaccine or drug anywhere in the world. Moderna's financial prospects hinged entirely on making that happen. In December 2018, the company's initial public offering on the stock market had set the record as the largest-ever in the biotech world, putting its estimated worth at $7.5 billion. Along with infectious diseases vaccines, the company had in its development pipeline potential mRNA-based therapies for rare genetic diseases and cancers. Yet in the ten years since its founding, Moderna had not brought a single product to market. Within the biotech community, doubts about Moderna's prospects grew with each passing year. "The only thing one can be sure of is the persuasiveness of their PowerPoint decks," an influential pharma blogger once sniped.

Bancel had heard it all. The criticism painted him as a brutal, unforgiving boss who cared only about the stock price — a portrait that didn't jibe with what the people who knew him best thought. What his detractors left out was that big ideas — the kind of ideas that could change the world, that could *save* the world — took big money, not to mention a person skilled enough and motivated enough to raise it. Bancel had always said that Moderna would either succeed spectacularly or fail completely. There was no in-between.

Once the CEO had settled into his cramped European-style hotel room in Davos, he fired off an e-mail to one man who could help with his mission: an American doctor named Richard Hatchett who was also at Davos.

Hatchett ran the Coalition for Epidemic Preparedness Innovations (CEPI), an organization headquartered in Norway that had been founded by the Bill and Melinda Gates Foundation in the wake of the market failures of the Ebola and Zika vaccines. There were other global organizations delivering lifesaving vaccines to countries that needed them, but CEPI's focus was on preparing for the next pandemic: disease X. Hatchett had been closely watching the coronavirus outbreak, and a member of his staff had already talked to Barney Graham. "Everything I had learned over two decades made me real anxious," Hatchett said. When he saw Bancel's e-mail in his in-box, he wrote back and agreed to meet immediately.

Hatchett had snow-white hair and a thin, bloodless face. He saw the viral world in terms of probabilities. During the meeting with Bancel, Hatchett expressed his alarm about the outbreak in Wuhan. No one could say yet how infectious this new, unnamed coronavirus was. Mathematical models of influenza, Hatchett knew, had shown that once a new strain has infected between thirty and fifty people, a pandemic became inevitable unless extraordinary steps were taken. If that held true for this new coronavirus, the world was already too late.

When Hatchett exhaled, Bancel updated him on the latest regarding Moderna's work to produce the first clinical batches of vaccine. Bancel recently had done a back-of-the-envelope calculation that had been shocking even to him. He discovered that Moderna could manufacture a pilot lot for a safety trial in humans using the small-scale equipment the company used for personalized cancer treatments. This brought the cost down from three to five million dollars to less than a million—a veritable bargain.

Even before the meeting, Hatchett and his team had wanted to be a part of the trial, and they swiftly organized a panel discussion to make a splash. On the afternoon of Thursday, January 23, inside the über-modern Davos Congress Centre, Hatchett announced that CEPI was giving Moderna a million dollars for the pilot lot that would be used in the NIH phase 1 trial. When it was Bancel's turn to speak, he offered a layperson's explanation of the company's technology in his pronounced French accent. He said it was like a piece of software that just needed to be updated for the latest outbreak.

Someone asked him how long he thought that would take.

"We don't have a timeline yet," Bancel said. "This has never been done before."

As Bancel and Hatchett filed out of the room, they checked their phones for the latest news from China. Wuhan's high-speed railway station had filled with travelers in puffy jackets and face masks vying for spots on the last trains out of the city. The airport had shut down, and authorities had begun blocking the major highways. Movie theaters were shuttered. Internet cafés were offline. Grocery-store shelves were emptying. Restrictions were now in place on more than fifty million residents across Hubei Province. The Lunar New Year celebration, China's biggest travel holiday, was officially canceled.

Two days after the press conference, Bancel sent an urgent message to Barney Graham, Tony Fauci, and John Mascola, the director of the Vaccine Research Center. "I have shortened my trip in Europe and am flying back to the U.S. tomorrow," he wrote. He would be coming straight to DC to strategize for the clinical trial and was already looking into securing additional government funding. "We need to make investments now to scale up the mRNA process," he wrote. "So lots to talk about. But on it."

The NIH-led teams were still working to grow the modified spike protein for their preclinical tests, which would include making sure that it was properly stabilized. Moderna's scientists, engineers, and technicians, meanwhile, were pressing forward on producing the mRNA vaccine itself. Every batch of vaccine had to be checked to ensure that the RNA was identical to the initial sequence and that the final vaccine was properly sterilized and safe. The fastest Moderna had ever rolled out a clinical-grade vaccine candidate was ten months, for the Zika vaccine. In the first days of the coronavirus project, one of the company's senior directors, Hamilton Bennett, asked a member of the team what its timeline would be. The response: one hundred and twenty days. "That's insane," she said. "Are you sure?" No. Actually, they wanted to produce the vaccine faster. The team members decided they would work overnight shifts seven days a week. "The virus doesn't take weekends," an engineer told Bennett. "So we can't either." Before long, the engineers had halved their timeline to sixty days. If the animal studies went well, a vaccine could be ready for the phase 1 clinical trial in March.

Bancel wasn't racing against the virus alone. Had he turned on the TV after his own press conference, he would have seen the face of Gregory Glenn, the chief science officer of another biotech company, Novavax, making the rounds at Bloomberg News and CNBC. During one interview, Glenn was seated at a desk inside his company's headquarters in Gaithersburg, Maryland, wearing a blazer and an oxford shirt with the top button open. "I have to confess it's an exciting time," Glenn said. "This is going to be a test case." Novavax, he said, was entering the coronavirus vaccine race.

If Moderna was the Apple of the vaccine world, an envy-inspiring company that was oozing with cool and bursting with capital, Novavax was Nokia, a has-been that was still selling something but you weren't entirely sure who was buying it. Novasomes, the company's trademarked encapsulation technology, were microscopic bubbles of fat developed as a timed-release mechanism for chicken vaccines. But they proved to be more valuable as a moisturizer in a skin-care line, Nova Skin Care. And thanks to their oh-so-creamy mouth-feel, novasomes ended up in low-fat Girl Scout cookies and in Richard Simmons's Slimmons, 97 percent fat-free cookies. Thirty years later, Novavax had dropped the nova-somes but was still trying, unsuccessfully, to get a human vaccine onto the market.

The company focused on protein-subunit vaccines. Unlike Moderna's mRNA vaccine, which delivered instructions to the body's cells on how to produce the spike protein, Novavax's vaccine delivered the spike protein itself. Protein-subunit vaccines were invented during the genetic-engineering revolution in the 1980s, and though they took longer to make than mRNA vaccines, the technique had a solid track record. You could grow virus proteins inside a variety of organisms, from *E. coli* bacteria to tobacco plants. Back in the 1980s, long before he became Novavax's head of vaccine development, Gale Smith had invented a technique that used one of the fastest — and strangest — platforms: caterpillar ovary cells. Whenever there was an outbreak of an infectious disease, from swine flu to Ebola, Novavax raced to produce a vaccine candidate in these cells, juicing its stock with a blitz of press releases. But every time, the company let its investors down. It let itself down. This time, Glenn and Smith hoped things would end differently.

To help in its efforts, Novavax turned to a company down the road in Baltimore, Emergent BioSolutions. Emergent mostly made drugs for other companies, and it had an entire manufacturing suite for pandemic preparedness that the company had been struggling to use. On the evening of Friday, January 24, Novavax's manufacturing guy, Brian Webb, called up Syed Husain, the head of Emergent's contract manufacturing business. "I have a crazy request for you," Webb began.

He told Husain that Greg Glenn and the team were trying to obtain a few million dollars from Richard Hatchett at CEPI, which had put out an open call for proposals. Webb wanted Emergent to provide an estimate on helping Novavax produce a clinical version of its vaccine candidate — even though Novavax still hadn't settled on the vaccine design. No problem, Husain said. His team would start roughing in some numbers and send something back sometime next week. "When do you need it?" he asked.

"Ideally, Monday morning," Webb said.

$\cdot$ $\cdot$ $\cdot$

Across the Atlantic, Sarah Gilbert was at work on a third class of vaccine. The first three cases of coronavirus had appeared in Europe, and *BBC Newsnight* wanted to hear what the Oxford professor was doing about it. An intense, and intensely private, woman in her late fifties with shoulder-length red hair and an inability to engage in small talk, Gilbert worked at the University of Oxford's Jenner Institute, named after the eighteenth-century smallpox-vaccine pioneer. The group had access to a small vaccine plant, part of Vaccitech, an offshoot of the Jenner Institute. Gilbert hated being in the limelight, but the head of her organization, an Irishman named Adrian Hill, was more of a showboat. If they were going to succeed at any of this, Hill certainly knew, they needed investment, and that required publicity.

Gilbert ended up giving the television interview with the oldest university in the English-speaking world serving as a backdrop. She told *Newsnight* that the Oxford group's preferred technology was known as a viral vector vaccine. Rather than delivering the spike gene on its own, as Moderna was doing, the Oxford group's method involved creating a Trojan horse by inserting the spike gene into a harmless virus. The virus would then enter a human cell and direct the cell's machinery to make the spike protein. Delivering the instructions via a live virus could potentially stimulate the immune system in ways that delivering the disembodied spike protein could not.

The first step with this technique was to identify the right virus to serve as the delivery vehicle. On the one hand, you wanted a virus that the human immune system had never seen before and therefore couldn't be stopped from infecting cells. On the other hand, the virus had to be handicapped in some way so that it couldn't become dangerous in itself. For an Ebola vaccine Gilbert had developed a few years earlier, her team had used a cold virus — an adenovirus — from chimpanzees that couldn't replicate in people. Because the Oxford researchers had brought that vaccine through late-stage clinical trials, they believed they would have an advantage in rapidly deploying it for the coronavirus. Unlike Novavax's and Moderna's vaccines, Oxford's vaccine, the researchers promised, could be delivered in just one dose, making it much easier to distribute quickly on a global scale. In the coming months, the handful of people working under Gilbert would grow to more than two hundred and fifty.

The coronavirus had, by now, made its first tentative incursion into the United States. On January 19, 2020, a thirty-five-year-old man had walked into an urgent care clinic in the Seattle area. He had a dry cough and was feeling nauseated. He had seen alerts about the coronavirus outbreak and grew concerned because he had recently visited his family in Wuhan. No one he knew in Wuhan had been sick, he said. He sat in a chair; his doctor tilted his head back slightly and inserted a long cotton swab horizontally into his nose until it reached the firm wall of the nasopharynx. Finally, the swab was withdrawn. A second swab was used on the back of his throat, likely causing him to gag. These two swabs were placed in a vial containing a gelatin and nutrient mixture to keep the viruses from deteriorating and sent to the Centers for Disease Control and Prevention for analysis.

The test came back positive late in the day on January 20. President Donald Trump was in Davos, and CNBC's reporters caught up with him. "We have it totally under control. It's one person coming in from China," he said, his arms outspread, palms up, with his trademark *What, me worry?* look. "It's going to be just fine." His impeachment, under way in the Senate, was foremost on his mind.

Three days later, inside the Dirksen Senate Office Building, Tony Fauci took a seat at a wooden table for a closed-door briefing with a couple of dozen senators. The head of the NIH's infectious diseases institute was joined by the leaders of two other divisions of the U.S. Department of Health and Human Services. The three men were about to be peppered with questions from the lawmakers.

To Fauci's right sat Robert Redfield, the virologist who led the Centers for Disease Control and Prevention. The agency's doctors and scientists were tasked with monitoring public health, conducting disease outbreak investigations, and keeping new pathogens from entering the country. A devout Catholic, Redfield looked like a medieval friar with his thick build and distinctive white chinstrap beard. Like Fauci, Redfield had been in the HIV field. In the 1990s, while at the Walter Reed

Army Medical Center, he had overseen a clinical trial of an AIDS vaccine, VaxSyn, that proved to be an embarrassment for him. He had tested the vaccine not as a preventive but as a therapeutic to boost the immune system of people already infected with HIV. Redfield ended up facing allegations of scientific misconduct after presenting an overly upbeat interpretation of the results at a conference, and while an investigation cleared him of the charges, the issue was not resolved entirely in the minds of his peers.

To Fauci's left was a man with a pink face wearing an ill-fitting gray suit. This was Robert Kadlec. Kadlec was not himself a scientist, though he had once been a pupil of Redfield's at Walter Reed. A former air force doctor, he was now the head of the Office of the Assistant Secretary for Preparedness and Response, the ASPR, a relatively obscure division within HHS. Ten thousand employees tracked diseases at the CDC and twenty thousand focused on biomedical research at the NIH, but Kadlec had just eight hundred staffers. His budget — a few billion dollars a year — was also dwarfed by those larger agencies. Nevertheless, Kadlec was likely to play a pivotal role in any outbreak once he received his marching orders. ASPR (pronounced "Asper") was like the Special Ops wing of HHS. It was supposed to be ready to respond to any health emergency, be it the aftermath of a hurricane, a nuclear weapons attack, or a flu pandemic.

The Senate session began, and lawmakers grilled the three men. How worried should the lawmakers be? Should the country consider locking down cities, as China was doing? Should they shut down international borders? What other measures needed to be taken?

Redfield told the senators he had spoken to George Gao, his Chinese counterpart, on January 3 but hadn't gotten much out of him. The United States was receiving more information from China than it had during the 2002 SARS outbreak, but that wasn't saying much. Now that the country had its first live virus samples, Redfield added, the CDC would be able to rapidly develop diagnostic tests and institute track-and-trace programs. He believed that the virus could be contained before it spread within U.S. communities.

Fauci spoke about the Vaccine Research Center's efforts, which included not only the vaccine collaboration with Moderna but also a partnership to screen the blood plasma from any recovered patients for spike antibodies and the white blood cells that secreted them. Those cells could potentially be cloned to manufacture monoclonal-antibody therapies, which could be rolled out to patients long before a vaccine was approved. He pointed out that the first SARS outbreak had resolved on its own without a vaccine or a therapy. The risk of the new coronavirus for Americans was still low, he said, and there was no need to close borders.

Kadlec, who was known on Capitol Hill as Dr. Bob, was growing more concerned by the day. He told the senators it was too early to tell for sure what the source of the outbreak was. It was most likely a natural spillover event, as the Chinese claimed, but it could also be an accidental leak from one of China's numerous virus laboratories. The Wuhan Institute of Virology, for instance, had been conducting research on bat coronaviruses for years. "Whenever you have a novel outbreak, it could be a bioweapon or it could be the *next* bioweapon," he warned darkly.

Kadlec went on to tell the senators about the potential weak spots the country would face if the outbreak expanded. The previous year, the ASPR had run a twelve-state war game, Crimson Contagion, of a hypothetical flu pandemic coming out of China. The scenario predicted 110 million infections, 7.7 million hospitalizations, and 586,000 deaths. To fight such a pandemic, Kadlec had concluded, HHS would require more than ten billion dollars in new funding, in part for vaccine manufacturing. One of the chief challenges would be ensuring tight coordination among the different branches of the federal government, which would receive competing requests for assistance from the states. Kadlec also told the senators how dependent the United States was on China for critical medical supplies, including drugs and N95 respirator masks, which can filter out viruses. The supply chain, he said, would be severely stressed should the outbreak continue.

Kadlec didn't divulge it at the time, but part of the reason he was so worried was that he had inside knowledge about what was going on in Wuhan. Two days earlier he had received a phone call from a gutsy American doctor who was at that very moment in the center of the hot zone.

3

NOT SANCTIONED, NOT AUTHORIZED

Wuhan had not been part of Michael Callahan's original itinerary. On January 17, he had stepped off a plane in the Chinese city of Nanjing after some thirty-odd hours in transit. The American doctor was looking forward to seeing his Chinese pals, a couple of veteran disease fighters he had met in Hong Kong during the 2002–2003 SARS outbreak. He had kept up the friendship by writing letters of recommendation to American universities for their students and children and by visiting China every couple of years for three to six weeks at a time for a collaboration on avian flu. "We're getting older, we're getting balder, but we're still the same people," he said.

As a wiry young ambulance medic and medical resident, Callahan had rescued people from plane crashes and mountain-climbing accidents. "Onesies," he called them—the isolated events that were mere stepping-stones to his first passion: disaster medicine. After he tired of tsunamis and earthquakes, he started suiting up for disease outbreaks. The first came in 1993, when he tagged along with the CDC's Viral Special Pathogens Branch on an investigation into a deadly hantavirus outbreak in the southwestern United States. Soon, he was jetting off to Asia and Africa; he always seemed to be the first guy on the ground at the scene of an outbreak. He once had to inject a crazed Ebola patient with a tranquilizer while trying not to poke a hole in his biocontainment suit. "You get infected, you die," he said.

In the early 2000s, Callahan's career took the first of many unexpected turns when the government sent him on a remarkable mission. His mandate was to form alliances with scientists at some of the most secretive bioweapons laboratories in Russia and the former Soviet states. "These guys ran the National Institute of Death for Russia, right?" Callahan said. His job was to harness their powers to fight disease rather than weaponize it. He remembered standing around a tundra lake with these killers, fishing for trout, barbecuing, and listening to Russian folk music as the scientists obsessed about how to get their kids into a good college. They'd hand him avian flu virus samples and give him advice on how to evade Russia's Federal Security Service, the dreaded FSB. "Don't leave tonight," they told him one time. The next time it was "You've got to leave Russia now!" When one of their scientists accidentally pricked herself with a syringe containing a deadly pathogen, Callahan parachuted in to assist in her treatment.

Later, Callahan joined the Defense Department, where he worked to accelerate vaccine and drug development and established an international disease-surveillance effort, enlisting the help of regional hospitals in a dozen countries, including Mexico, Nigeria, and Indonesia. Now fifty-seven, Callahan had returned to civilian life. His hair was indeed thinning and his days weren't as action-packed as they once were, but he hadn't slowed down.

Nanjing, which is situated on the banks of the Yangtze River, is China's ancient southern capital and a modern-day center for science. It was a pleasant enough place to go to escape winter weather in Boston, where Callahan held a part-time position at Harvard's Massachusetts General Hospital. When Callahan set down his bag in the lobby of the Westin Hotel and was handed the key card to his room, he had to smirk. *There's four hundred rooms in this hotel, and I get the same room every time?* he thought. It was a fine room. Clean bathroom, firm mattress. It was also a tell. Ever since Chinese hackers stole a database containing information about his high-level security clearance, Callahan knew that someone might be watching his every move. "I'm not that good-looking of a guy, but you'd think I was Brad Pitt when I go down and get a beer," he said. "Honeypots. But, you know, we get training for that."

For weeks, Callahan had been following the chatter coming out of Wuhan, some three hundred miles upriver from Nanjing. Shortly after Callahan arrived, his Chinese doctor colleagues heeded a country-wide alert to assist with the outbreak. Callahan could feel his own chest tighten at the thought of another hot zone. There was something about that hypervigilant state that made him feel most alive. If he went to Wuhan, Callahan knew he couldn't worry his wife by telling her about his plan. He had to be careful about telling anyone. He didn't have official permission to travel there, after all. "It was not sanctioned, not authorized," he said.

He went to Wuhan anyway and hunkered down in another hotel, waiting to get the word from his friends. "They had to check in to make sure things were safe for me." On January 22, Callahan slipped on medical scrubs and donned an N95 mask and a pair of goggles to pass through the entrance of the Wuhan Central Hospital, a boot-shaped glass building rising up from the city's empty streets. There, his colleagues registered him as a "guest clinical educator," a title that would allow him into the wards as an observer. The next day, the city locked down. Callahan had just made it into the white-hot center of the outbreak.

He soon saw the familiar look of terror and confusion in the eyes of Wuhan's infected patients as their shallow breathing quickened and their desperation grew. Perhaps they had inhaled a few hundred virus particles in the air. The phalanx of invaders then touched down on the hairs in their nostrils or deep in the sinuses and upper airways. After a couple of days, the victim felt the itch in the back of the throat, the first sign that something was amiss. A few cells here and there were dying, releasing chemical SOS signals as they perished. In these first stages of a coronavirus infection, the immune system launches a generic defense, ramping up mucus production to flood out the invader and create a physical barrier between the pathogen and the cells. It also amplifies those SOS signals and deploys white blood cells, the foot soldiers of the immune system. This is when the coughing and sneezing begins, and maybe a little fever. Having never been exposed to this virus before, a person's body had no targeted means to fight it.

Gradually, more specialized white blood cells arrive on the scene, a process that can take longer in older patients and others with weakened immune systems. During this ramp-up, the body is homing in on specific weapons to fight the virus: antibodies. Antibodies are grabby,

Y-shaped proteins that cling to other molecules. About one-fifth of the weight of your blood serum is made up of this whitish gunk. Through combinatorial magic, your body makes as many types of antibodies as there are stars in the Andromeda galaxy—one thousand billion. Some of them, by chance, will bind to the coronavirus, but only a vanishingly small number will disable an individual virus before it can infect a new cell. Afterward, these antibody-decorated viruses are swept away, sucked up by the body's house cleaners.

While the antibody response is still escalating, the coronavirus can largely multiply with abandon. This is its moment of opportunity. It chews through human cells one at a time, turning them into infective slime. That slime oozes down the victim's throat, lodging itself in the lungs, where infection is harder to clear. The body launches its riskiest line of defense—it sends in its assassins the killer T cells, which seek out infected cells and trigger their self-destruct buttons. As this war zone inside the lungs heats up, collateral damage becomes inevitable. Half of the immune system ends up fighting the other half. Red blood cells burst and disgorge their hemoglobin, an iron-rich molecule that wreaks havoc in the lungs, like a grenade mistakenly dropped in the trenches. As their injured lungs fill with a toxic jelly of dying cells, the patients feel like they are drowning. This desperation happens long before the patients actually are drowning, and it is the doctors' job to keep the patients in the game, fighting the disease with their own lung power, aided by supplemental oxygen, steroids, and pain relievers. When they got to a point when the drugs were no longer working, when the oxygen wasn't helping, when the pulse oximeter readings dropped to 70, 60, 50 . . . there was only one thing Callahan's colleagues could do to keep them alive: anesthetize them and place them on ventilators.

Those ventilators sat on wheeled pedestals so they could be quickly moved on to the next patient when the current one no longer needed it. Except what Callahan was seeing was that patients were being put on ventilators faster than they were coming off them. Caseloads were rising from the dozens to the hundreds. Every time a new breathing tube was inserted into someone's throat, infectious clouds of virus particles filled the air. Callahan helped his colleagues set up so-called laminar-flow rooms as he had during SARS, positioning fans on the windows to suck contaminated air out. The hospital was running out of rooms, running out of ventilators. Even at this early stage, he could see that this was going to be worse than SARS.

Callahan was witnessing only the most dramatic and immediate impacts of severe infection, the way it stressed the lungs, the heart, and the kidneys. But when you push the immune system so hard, when your body is repeatedly detonating these grenades, not all of that damage can be repaired. Those patients lucky enough to emerge from the intensive care unit alive were not going to pop back into their normal lives the moment they were released. They would continue to have trouble breathing. They would have a persistent cough that might never go away. Joint and chest pain too. Some would suffer from a sort of brain fog that made it difficult to concentrate, leaving them in a permanent funk; depression would settle on them. Others would have their fevers and chills return periodically, along with a racing heartbeat or other arrhythmias. Even less severe coronavirus infections, Callahan knew, had the potential to scramble the normal functioning of the human body, causing people to develop rashes or lose their hair.

All told, Callahan spent almost a week on the ground helping his colleagues keep the hospital functioning, learning about the virus's toll on the human body, and taking note of what drugs doctors were throwing at the virus. Chinese officials were planning to tighten Wuhan's quarantine measures, banning residents even from stepping out to buy food. Callahan slipped across the river by boat — "the black-market way" — and returned to Nanjing, where he and his colleagues had a video link with the ICU units in two hospitals in Wuhan and could provide advice and track patient outcomes. Callahan knew he needed to report what he was seeing to his friends in the U.S. government.

The headquarters of the Department of Health and Human Services are located inside the Hubert H. Humphrey Building, an eight-story brutalist concrete building straddling both a freeway tunnel that burrows underneath the capital and a major sewer line that carries all the crap emerging from it. The department controlled a budget of more than one trillion dollars, an amount larger than the entire military, and its dominion included the NIH, the CDC, the FDA, the Center for Medicare and

Medicaid (CMS), and a number of lesser-known divisions, one of which was the Office of the Assistant Secretary for Preparedness and Response.

Up on the sixth floor, Bob Kadlec's grim, windowless office was a messy affair. Almost every inch of desk and table space was covered with precarious stacks of papers, binders, and manila folders; one small area held a bowl of candy with an American flag planted in it.

On Tuesday, January 28, both Kadlec and Brett Giroir, a former mentor of Callahan's and the HHS assistant secretary of health, received an e-mail from Callahan. He told them he had seen data showing that the incidence of the disease was four times that being reported to the WHO. The Chinese had 23,000 people under daily observation with confirmed infectious contacts. Of the 277 closely monitored patients he was following, twenty-two had been released and one had died. The virus stayed active in a person's body for about nine days, which he believed posed "a major challenge." One of the most wily aspects of coronaviruses was that some people were seriously affected and others not so much. Viruses like that were much more difficult to contain because they so easily flew under the radar. Asymptomatic and minimally ill patients, Callahan wrote, "will propagate virus into distant communities." The new coronavirus was a serious disease, but its fatality rate was anyone's guess. It wasn't Ebola, which kills an average of half the people it infects, but it was worse than the seasonal flu, which kills 0.1 percent of infected people over the age of sixty-five. He warned Kadlec and Giroir that the monoclonal-antibody therapies that had worked in the first SARS outbreak were proving useless against the new coronavirus.

Kadlec had known Callahan for twenty years, and the infectious diseases doc had likely now seen more patients infected with the new coronavirus than any other American doctor had. But what he was saying didn't comport with the early reassurances Kadlec had gotten from the CDC or with the near vacuum of information coming from the intelligence community. "The two most high-confidence sources of information I have can't tell me shit," Kadlec said. "No one had a grasp on what was going on." Over the previous two days, the number of confirmed cases reported by the Chinese in Hubei Province had jumped by 65 percent, from 2,744 to 4,515, confirming that Callahan knew exactly what he was talking about.

Dr. Bob began to envision the pandemonium that could potentially unfurl inside American borders, when the crisis would become his responsibility. After a nine a.m. meeting with some of the department's leadership, he pulled Brett Giroir aside and asked him what he thought of Callahan's message. Giroir was a fierce man with a buzz cut and a square jaw who dutifully wore the navy uniform of the U.S. Public Health Service Commissioned Corps, of which he was the top doctor. He agreed that it was all very scary and that they needed to convince the rest of the team at HHS, including the secretary of health, Alex Azar, to ramp up the nation's response to the virus.

Within days, Kadlec had signed Callahan up for a six-month stint helping ASPR to provide virus intelligence by way of his Chinese connections and advise on how to respond to virus inside the United States. "His physical location and meetings will be determined by the ASPR," read his statement of work. "He will be expected to respond on ICS [Incident Command System] timelines (24/7 mobile access: 2hr to airport; autonomous resources). He will not [be] official U.S. Government, will not represent U.S. Government opinions, will not communicate to media or social media and will presume all information is SBUC/NOFORN."* The wild man in Wuhan, in other words, would be on call for Americans twenty-four hours a day.

---

* Sensitive but unclassified / not releasable to foreign nationals

4

OUR BUGS AND GAS GUY

Mission control for Health and Human Services is known as the Secretary's Operations Center, or the SOC. It's located down the hall from Bob Kadlec's office in a large room with a bullpen at the center and computer stations in rows on either side. Screens on three walls display data and video feeds for whatever crisis the agency is currently managing. In late January 2020, the SOC came alive as it shifted from a virtual footprint, known as level three activation, to the more critical level two, where a staff made up of emergency responders, ex-military types, and medical experts were called in to coordinate the coronavirus response among different state, local, and federal agencies.

All through the night of Tuesday, January 28, Kadlec's team in the SOC was making calls and anxiously watching a blip on a screen heading toward the Alaskan coastline. This was a chartered Boeing 747 — Kalitta Air flight no. 317 — the first State Department repatriation flight carrying more than two hundred Americans who had been trapped in Wuhan after the lockdown. It was due to land at Ted Stevens Anchorage International Airport just before 9:30 p.m. Alaska time — four time zones to the west. The temperature there was well below freezing. Inside the SOC, the team scrambled to figure out where to shelter those two hundred Americans.

The slogan of the Office of the Assistant Secretary for Preparedness and Response is "Public Health and Medical Emergency Support for a

Nation Prepared," evoking a paranoid "duck-and-cover" campaign from the Cold War era. Not only did Kadlec's office respond to public health emergencies like this one winging its way from Wuhan, it was also charged with preparing for whatever might come next by ushering new drugs and vaccines onto the market. In one of the agency's informational videos from 2019, Kadlec looks like a TV weatherman. He's standing inside the SOC wearing a beige suit and a pale yellow tie. His gray hair is swept to the side. His frameless glasses are perched on his shiny beak. He has a quavering voice that often veers into the higher registers. "If there's a mass casualty event," he says, "something as simple as quickly knowing how many hospital beds are available from all the hospitals in an area can often take way too long." Then he unconvincingly repeats the refrain "There has to be a better way," and you can imagine his employees shaking their heads. Kadlec wasn't much of a showman.

What he was was a spymaster. In the early days of the outbreak, he listened in on the conversations of a group of biosecurity experts who had gotten to know one another during the George W. Bush years. Kadlec had dubbed them the Wolverines, a term ripped from the 1984 movie *Red Dawn*, about insurgent farmers defending the United States from a Russian invasion. The Wolverines were all close friends with Michael Callahan, and each member had his own special expertise. For instance, Richard Hatchett, the head of CEPI in Norway, and Carter Mecher, an adviser to the Department of Veterans Affairs, had conducted research in the 2000s that demonstrated the potential value of low-tech measures to fight disease outbreaks in the absence of any drugs or vaccines. They had found that during the 1918 influenza pandemic, social-distancing measures, such as the closure of schools and churches and advising the public to "avoid crowds," lowered death rates by half in the cities where they were put in place.

On an e-mail list created by the Wolverines, Hatchett wrote that he was haunted by the specter of swine flu, which had spurred a vaccine race but was ultimately not as deadly as anticipated. "The crying-wolf scenario," Hatchett called it. Beyond the small stake in Moderna and two other companies, he was struggling to decide how much investment CEPI should put into vaccine development and how to court donors fearful of "political embarrassment" and "accusations of mismanagement of funds." He wrote, "Grappling with both horns of the dilemma here. Would welcome you wrapping your brain around how to proceed in the most prudent way. . . ."

James Lawler chimed in next. He wasn't looking at this as a crying-wolf scenario. Like Callahan, Lawler was a hot-zone guy, a physician at the University of Nebraska Medical Center in Omaha, which had one of the quarantine units suitable for the nastiest diseases on the planet. "Great understatements in history," his tongue-in-cheek message began. Pompeii, he wrote was "a bit of a dust storm." Hiroshima was "a bad summer heat wave." As for Wuhan, it was "just a bad flu season."

Matthew Hepburn, an army doctor who ran a vaccine program at the Defense Department, replied that he too was on the horns of a dilemma. "Team," he wrote, "I am dealing with a very similar scenario, in terms of not trying to overreact and damage credibility. My argument is that we should treat this as the next pandemic for now, and we can always scale back if the outbreak dissipates, or is not as severe."

For Kadlec, the mitigation playbook in response to a pandemic came down to what he called the four Ss. The first is "shield the vulnerable," such as the elderly and the infirm. Second is "shelter the susceptible," which means breaking the chain of infection in the wider community through social-distancing measures or mandatory lockdowns. Third on the list is "save the sick" with whatever drugs and equipment you have available. Last is "sustain supplies," such as personal protective equipment for doctors.

Although members of his team were working on vaccines, that wasn't high up on Kadlec's triage protocol for the coronavirus. That was part of *preparedness,* not *response.* By the time disease X arrived, it was usually too late for preparedness. Vaccines took years to develop and proposed candidates frequently failed. Kadlec had to focus on the thing that was right in front of him: a looming disaster. He controlled the Strategic National Stockpile, a system of warehouses filled with eight billion dollars' worth of drugs, vaccines, and medical supplies, but it was meant only to back up individual states and localities in the event of crisis. No stockpile was large enough for a sustained nationwide epidemic, and the

nation was potentially entering one with a stockpile that had been neglected for years.

More than twenty years earlier, Kadlec had helped draft the law that created the ASPR, but he never expected that one day he'd lead it. He grew
up on Long Island, in a community east of John F. Kennedy Airport.
His parents, who were of Czech descent, had been imprisoned by the
Nazis during World War II. They wanted nothing to do with war, but
Kadlec's imagination was captured by films about fighter pilots, including *Twelve O'Clock High* with Gregory Peck and *Strategic Air Command*
starring Jimmy Stewart. He entered the U.S. Air Force Academy in Colorado Springs in the 1970s, where he promptly had his head shaved and
his ass kicked. He took survival and assault courses and had to eat his
meals sitting at attention, heels together, back straight.

His goal was to become a flight surgeon, the evocative title given
to any physician taking care of aircrews. In 1984, he was assigned to
the First Special Operations Wing at Hurlburt Field, near Pensacola
on Florida's Gulf Coast. Over the four years Kadlec spent on the base,
he became known as an innovator. Most flight surgeons focused on
aviation medicine — g-forces and all that — but Kadlec was interested in
operational medicine, which was about treating soldiers in the field. "It
was kind of a backwater," he said. He worked with the Army Rangers
to turn the rear of a C-130 Hercules plane into a mini-hospital, which
earned him the title of Flight Surgeon of the Year.

It was during that time that he met Joshua Lederberg, the Nobel
Prize–winning microbiologist who later wrote that "the single biggest
threat to man's continued dominance on this planet is the virus." Kadlec
assisted Lederberg on an evaluation of the base's ability to respond to
biological weapons and became fascinated by the topic. After earning
a master's degree in tropical medicine at Walter Reed Army Medical
Center, he was recruited for Special Operations training in Fort Bragg,
North Carolina. Almost as soon as Kadlec started work, on August 2,
1990, Saddam Hussein invaded Kuwait. Wayne Downing, commanding general of the Joint Special Operations Command, turned to Kadlec
and said, "Bobby, you're our bugs and gas guy." Iraq had an active biological weapons program, including one notorious facility disguised as

a chicken-feed plant. Downing sent Kadlec over as the physician of a counterterrorism unit on the Saudi Arabian border. Most of his time was spent watching *Animal House* and Monty Python movies on repeat until he was called to pull dead soldiers out of a helicopter and put them into body bags.

Kadlec went on to launch the U.S. government's first-ever bioweapons vaccination program, immunizing over a thousand Navy SEALs, Army Rangers, and Delta Force soldiers against anthrax and botulinum toxin, two agents that had been stockpiled in Iraq. Then, for about three years, he was detailed to the Central Intelligence Agency's clandestine service, debriefing defectors and conducting field operations related to bioweapons.

In a book chapter he wrote in 1998, Kadlec described how a few kilograms of anthrax spores disseminated around New York City could kill four hundred thousand people in a matter of days. He considered how the United States might respond to such an event and concluded the country would pretty much be screwed. The CDC was focused on public health surveillance, tracking obesity rates and measles outbreaks. Researchers in the Department of Defense and the NIH studied infectious diseases, though neither of those institutions had sufficient means or the mission to bring its discoveries to the market. As for FEMA, the Federal Emergency Management Agency, it could move people and supplies around the country at a moment's notice, but it wasn't a health organization. Who knew how FEMA would fare in the face of a biological weapons attack?

The conclusion to this thought experiment, Kadlec wrote, was that the nation needed "a capable, practiced, and coordinated response mechanism," which would involve bringing together certain health-related functions of various agencies. He envisioned a vast stockpile of "necessary antibiotics, immunoglobulins, and vaccines" that were "readily available to administer within hours." He also thought the government needed to manufacture certain vaccines itself, since there wasn't much of a market for an anthrax vaccine. At least, not until disaster struck.

Kadlec offered his bioterror shtick to anyone who would listen. Few did. Then 9/11 happened, followed by the anthrax attacks—a series of anonymous letters sent to media organizations and Democratic senators that killed five people and sickened seventeen others. Suddenly, Kadlec was in high demand. He was drafted to work on President Bush's Homeland Security Council, which was where he and Tony Fauci met. "We've been through the wars together," Fauci said. "I would always tell him that I worry more about a natural occurrence of an outbreak rather than somebody deliberately releasing something." But the two saw a synergy in their interests. Kadlec helped Fauci's institute at NIH secure a piece of the Homeland Security budget in order to develop new vaccines against anthrax and smallpox.*

Dr. Bob found allies on Capitol Hill. Foremost among them was Senator Richard Burr of North Carolina, an oddball Republican who drove Kadlec around in his 1973 Volkswagen Thing with holes in the floorboards and bumper stickers touting conservative causes. Kadlec worked as the staff director for Burr's bioterrorism subcommittee. In 2004, Congress passed the Project Bioshield Act, which provided six billion dollars in funding for the government to purchase and stockpile countermeasures against anthrax, smallpox, and other threats. After FEMA utterly bungled its response to Hurricane Katrina in 2005, Kadlec helped Burr draft the law that created ASPR and its drug and vaccine development arm, the Biomedical Advanced Research and Development Authority, or BARDA.

When Kadlec left government in 2009, he tried his hand at private consulting and lobbying, but he didn't have the knack for it. "I just didn't have the fire in my belly to make money or be a salesman," he said. "Much to the disappointment of my wife." Part of the problem was that the former spy didn't spend as much time developing business leads as he did sharing intelligence leads with his handler at the CIA. In late 2013, he went back to work with Burr on the Senate Intelligence Committee with every intention of retiring after two years. When President Trump was elected in 2016, Burr asked Kadlec if he'd take the job as assistant secretary for preparedness and response.

I'd do it for free, Kadlec thought. His wife, Ann, an anesthesiologist,

---

* Developing a new smallpox vaccine became one of Barney Graham's first projects at the NIAID.

wasn't happy about this turn of events. Their twin daughters, Margaret and Samantha, were then in high school. When he told them about the potential appointment, he got an earful. Margaret, the more outspoken of the two, had come out as gay a couple of years earlier, writing the message to her parents in caramel sauce on a bowl of ice cream. "I don't want you working for that fucking man!" she yelled at her father in the kitchen one night. Trump had won the election on a platform of hate, she said. His family voted three to one against his taking the job.

"They are calling me, and I've got to answer the call," Kadlec responded. "If I don't take this job, some other schmuck will."

Kadlec considered himself a Colin Powell Republican, a national security hawk with a social conscience, but he wasn't driven by ideology. His thoughts were on arcane matters of health and security. When he arrived at the ASPR, he handed out business cards from his Pentagon days listing "Kadlec's Rules of Military Medicine." Rule no. 4: "Pride, complacency, and parochialism will kill more troops than any bullet, bomb, or missile." John Redd, a public health–minded doctor who led ASPR's medical missions, stiffened whenever Kadlec spoke in military terms, but the two grew to like and admire each other. Kadlec was always eager to hear stories of Redd's work outside the Beltway. "He's interested in what people in DC call 'the field,'" Redd said.

Kadlec didn't know what to make of President Trump and didn't expect to interact with him much. The functioning of a gargantuan organization like HHS depended on him reporting not to the president, but to the secretary of health — Tom Price at the time. Eight weeks after Kadlec started the job, Price was forced out and Alex Azar took over. Kadlec, meanwhile, had to deal with the one-two punch of Hurricane Harvey unleashing floods in Texas in August and Hurricane Maria leveling Puerto Rico in September. He flew with the president to Texas and followed him around as he glad-handed local officials. The next year, ASPR was saddled with solving a disaster of Trump's own making: Kadlec's team had to try to reunite the families of thousands of immigrants the administration had separated at the southern border.

One of the things Kadlec discovered early on was that his agency had deviated from the mission he had envisioned for it. His disaster medical-assistance teams (DMATs) were down to 2,200 people although

Congress had conveyed a budget allowing for 4,500, and they were no longer being trained in managing infectious diseases. Kadlec also became frustrated with the CDC during Hurricane Maria, when the agency had taken hours to respond to ASPR's requests for access to the Strategic National Stockpile, which the CDC controlled at the time. Equally troubling to Kadlec was the fact that the CDC had been using some of the stockpile appropriations to fund other projects, which contributed to the White House Office of Management and Budget placing it under ASPR's control.

Soon, Kadlec became known at the tightfisted budget office for requesting large sums of money to bulk up the stockpile. *A pandemic? Yeah, right,* came the reply from the OMB year after year. White House staff would scrunch up their brows as the long-winded Kadlec droned on about George Washington inoculating his troops against smallpox. Kadlec struggled to grasp why the administration didn't share his concern about biological threats. Perhaps these people lacked the imagination to grasp that even a low-probability scenario was an eventuality that the nation had to prepare for. Or perhaps they were just playing the short game, as they were in their denial of climate change. They figured that, more likely than not, Trump would no longer be in office by the time a once-in-a-century calamity struck.

One of the unwritten rules of being a political appointee is you cannot lobby Congress for an appropriation without the administration's approval. But if your supporters in Congress happen to ask about your agency's shortfalls, there's nothing stopping you from sharing them. During a hearing on ASPR's budget in 2019, Senator Burr lobbed a softball query to his old friend. "Dr. Bob," he said, "my first question is simple. Are we prepared for public health threats we face?"

"We have a long way to go," Kadlec responded. He feared the country would not be prepared for a terrorist attack or a pandemic coming from Asia. For instance, during the 2009 swine flu pandemic, the Obama administration had distributed three-quarters of the N95 masks in the stockpile and never replenished the supply. Kadlec had been forced to do some creative accounting since he'd taken over and he wanted Congress to double his budget. "We are operating with about half an aircraft carrier of resources to . . . basically protect three hundred and twenty million people," he told Burr. He received some, but not all, of the funding he felt necessary.

Not long after that, during a talk Kadlec gave at an industry event in the fall of 2019, he repeated his desire for more funding. "We're waiting for the next big thing to happen because you know it will," he told the crowd. "It could be a coronavirus — for which we don't have countermeasures."

Shortly before four p.m. on Friday, January 31, 2020, Secretary Alex Azar, a lawyer and former pharmaceutical executive, stepped up to the podium at the front of the White House Briefing Room, sporting a beard he had grown over the Christmas holiday. Azar was the chair of the newly created White House Coronavirus Task Force, which would keep the public informed about the state of the threat it was facing and coordinate the government's response to the outbreak across multiple divisions, including the State Department and Homeland Security.

It was a tall order for the health secretary, whose standing with the White House, not to mention his own subordinates, had become shaky. The son of a respected eye surgeon, Azar was an establishment Republican whose disciplined management style, with a focus on planning and process, was repeatedly challenged by the chaos of the Trump White House. Azar had been the deputy health secretary under George W. Bush during the first SARS outbreak and he and Kadlec had gotten to know each other then. In contrast to the dispassionate airman, however, Azar could get worked up. Grudges — he had a few of those. In a time of crisis, he would need to set all that aside to foster unity across the balkanized health department.

Azar initially appointed Tony Fauci of the NIH and CDC director Robert Redfield to the task force. But the reverence for science that Azar's father had impressed upon him would be counterbalanced by officials the White House selected, including, to Azar's great dismay, Joe Grogan, a radical conservative from Albany. A former budget official, Grogan ran the powerful domestic policy shop on the third floor of the West Wing and was hellbent on sabotaging anything with Azar's name on it. Grogan had crushed Azar's early attempt to bring transparency to drug pricing and had outmaneuvered Azar in a battle over the use of tissue obtained from abortions in federal research. Although Azar, a deeply religious man, opposed abortion, he had sided with Fauci and other NIH scientists in arguing for a less restrictive policy.

Standing in front of the task force, Azar arched his eyebrows as he spoke, enunciating the two halves of that still unfamiliar word — *"co-rona . . . virus"* — like a professor on the first day of class. He announced that the president would wield his authority to ban most travelers from China. "Since taking office, President Trump has been clear: His top priority is the safety of the American people," Azar said, transparently flattering his boss as he so often did. A fourteen-day quarantine for U.S. citizens returning from China would be required. Foreign nationals would be denied entry entirely. The China hawks in the administration had been advocating for a move like this for several days, but Azar had only agreed to it once he had the buy-in of the two agency heads at his side.

The first person Azar invited to speak to reporters was Redfield, who was sporting a navy-blue three-piece suit. In those early days, the CDC was the point agency coordinating the country's coronavirus response. "This is a serious health situation in China," Redfield said, "but I want to emphasize that the risk to the American public currently is low." He provided the latest rundown of statistics: 6 confirmed cases in the United States and 191 others under investigation. Redfield believed the CDC could still contain the coronavirus and keep it from sparking an epidemic inside the U.S. The agency was now screening arriving passengers at twenty airports by asking them whether they had certain symptoms and taking their temperatures.

Kadlec watched Redfield warily from the flanks. Although the CDC provided a trusted channel for communication around public health, he knew the agency could be slow-moving and insular, filled with office-bound number crunchers. *The CDC staff write reports; they don't respond to emergencies,* he thought. *That's what the ASPR was created for, damn it.* Until Redfield's scientists conceded that the coronavirus could not be contained, though, Kadlec's ASPR would remain in a supporting role. He was just the grunt, after all.

Behind closed doors, however, members of these two branches of Azar's empire had already been bickering over how to handle the two hundred American evacuees that had arrived from Wuhan. The CDC was developing quarantine guidance to describe how to isolate an infected patient in a hospital. It was silent on what to do with a 747 full of passengers. When it came to implementing quarantines, the CDC left that to the states, which were throwing their hands up in the air. As the ASPR staff was belatedly roped in to find appropriate lodging for those passengers, the CDC's dutiful quarantine officer Martin Cetron stuck to the book, insisting at every turn that their plans didn't meet the standards. Kadlec eventually finagled rooms for the evacuees at the March Air Reserve Base in Southern California.

Toward the end of the January 31 press briefing, a reporter asked Azar a question: "What measures do you have, sort of, in the toolkit if this gets worse?" Azar directed this query to Kadlec. It was Kadlec's first major public appearance since the start of the outbreak, but he didn't have either the scientific authority of Redfield or the pugnacious quotability of Fauci. "My role in this is really precautionary at this stage," he explained. "We're really taking a very — I would say cautious — but deliberate and methodical approach to ensure that, come what may, we're prepared."

5

STABILIZING THE SPIKE

In the final week of January, Kizzmekia Corbett lost seven pounds and slept, in her estimation, twenty hours total. She'd arrive at the NIH's Vaccine Research Center early in the mornings and would leave long after the sun had set, bleary-eyed and exhausted. Corbett, who had known Barney Graham for almost fifteen years, had made a second home for herself in his lab amid the benchtop centrifuges, gene sequencers, and specialty dishwashers designed to sterilize the Erlenmeyer flasks, graduated cylinders, and the rest of the scientific glassware the team cycled through every day. When Graham walked into the lab one time and heard Young Jeezy's *Let's Get It: Thug Motivation 101* blasting on computer speakers, he paused for a long second, swiveled around, and returned to his office without complaint. That, Corbett cracked on Twitter, was a boss attuned to "cultural sensitivity."

One of the students working under Corbett on the Moderna vaccine project was a woman named Olubukola "Olu" Abiona. She was charged with growing the coronavirus spike protein in cell cultures. While Moderna's mRNA vaccine was designed to generate spikes in the human body, researchers needed to produce their own spikes in laboratory cultures. The spike would be a key ingredient in an assay to determine if the vaccine produced the right immune response in mice, monkeys, and people. Researchers would also send the proteins to collaborators who would make diagnostic tests and monoclonal-antibody therapies. Normally, you would begin by growing a small amount of protein in a one-liter flask, but the team needed to go big so they would be ready when Moderna's vaccine candidate arrived.

Corbett's phone dinged. She saw a text from Abiona. "Ladies and gentlemen of the CoV team," she wrote. She and a lab mate had con- centrated and filtered four liters of the nutrient-rich broth, producing a whopping twenty-two milligrams of protein. The weight of a hundred snowflakes. It was enough to last a couple of weeks. "Yasssss bitchhhhh," Abiona opined.

Corbett ran over and gave her hug. She thought Yasssss bitchhhhh would make a great name for their vaccine. ("Unfortunately, I don't think [Moderna] will go for that," she later wrote. "We'll just call it YB on internal in vitro assays.") A few minutes later, Graham called Corbett asking for an update. As close as they were, she was always nervous to be put on the spot by him. When she saw his number, her "hair turned into an Afro from the sweat," she later tweeted. She told him the news and he said she could call the vaccine whatever she wanted.

Graham and Corbett were on edge on the night of January 30, wait- ing for Daniel Wrapp to produce images of the spike protein Jason McLellan's lab had grown in Texas. No one knew if it had been properly stabilized in the right conformation to be an effective vaccine. If the pro- tein's shape changed too easily, the right kind of antibodies would not be able to find their targets, and all of the researchers' hard work would go to waste. Those twenty-two milligrams of protein would be flushed down the toilet. Moderna might have to scrap the vaccine it had begun manufacturing.

Late that afternoon, Wrapp had taken vials filled with spike pro- tein over to the building housing the molecular microscope, a machine called the Krios — "sort of like the monolith from *2001: Space Odyssey*" is how Wrapp describes it. As he had rehearsed so many times before, he pipetted the protein solution onto a fine-mesh grid, carefully blotted off extra moisture, and then froze the solution. He then inserted the tray into the machine, which blasted the protein with a high-powered beam of electrons.

"There's a big scientific moment of truth coming up in about an hour and I'm a little nervous," he wrote his girlfriend that night. Later, he sent her a black-and-white image that looked like a photographic negative of the night sky—the stars black, the sky white. But they weren't exactly stars. "Those little blobs are the protein," he wrote her. "Yay!" she replied. After zooming in on the best images, the team saw that many of those spike proteins were locked in the shape that had the best chance of working as a vaccine.

When Graham heard the news, he felt a sense of tempered relief. The first hurdle had been crossed, but he had been through this process many times before and knew they were still a long way from the finish line.

Back in 2000, when Tony Fauci asked Graham to come to the Vaccine Research Center to run clinical trials, Graham was a rising star in the HIV/AIDS world. Graham had known Fauci since the 1980s, when they both worked on the first-ever clinical trial for an HIV vaccine, VaxSyn, which Robert Redfield later pushed as therapeutic. Though the initial VaxSyn trial was unsuccessful, it represented a major milestone in the field. Most scientists wouldn't have had to think twice about accepting a plum job at the NIH that allowed them to direct some of the most important research in the field and help stop a slow-moving pandemic that had killed nearly half a million Americans. Graham, however, hesitated.

His issue was that, while the VRC was focused on HIV, he had become intrigued by a diabolical pathogen that most people had never heard of: respiratory syncytial virus (RSV). First identified in 1956, three years after Jonas Salk announced his polio vaccine, RSV became a major cause of death among newborns. Scientists immediately went to work on a vaccine. During a clinical trial at the Children's Hospital in Washington, DC, in the late 1960s, infants and toddlers who were given an experimental vaccine and later contracted the virus ended up with a more severe form of the disease. Eighteen of them were hospitalized. Two died. It was a devastating turn of events and added to a growing recognition that, despite the early breakthroughs in the field, vaccinology was finally hitting a wall. How, Graham wondered, had a vaccine made a disease *worse*?

Graham felt that if he could unlock that answer, so much else about the human immune system would be revealed as well. He told Fauci he would come to the VRC only if he could continue his RSV work on the side. "We wanted him so badly we made a deal," Fauci said. It was a decision that would help make the VRC the tip of the spear in the country's ability to prepare for disease X.

Graham traces his own interest in science back to his life on his family's farm. He was born in Kansas City, but his father, a dentist by trade, moved the family south to the small town of Paola to raise horses, cattle, and hogs. When Graham wasn't playing basketball or studying, he and his brothers helped out around the operation, which often meant dealing with broken equipment. Though Graham knew how to repair a tractor and had graduated valedictorian of his tiny high school's class, he was behind many of his peers academically when he enrolled at Rice University in Houston in 1971. During one of his first physics tests, the answer to a problem was a minuscule number — a decimal point followed by a whole lot of zeros. It was so small that Graham figured he might as well write 0. He got it wrong. His roommate, a city slicker named Bill Gruber, found that gaffe far more humorous than Graham did. Graham told him he wasn't going to be fooled like that again.

Graham returned to Kansas for medical school. His parents were taken aback when he introduced them to the fellow medical student he planned to marry: Cynthia Turner, a Black single mother. "They were initially pretty hesitant and concerned by what it would mean for my family," Graham said. "After a few months, there was a coming together of minds."

Cynthia had reintroduced Graham to his Christian roots. He was raised as a Presbyterian but had strayed in his youth, focusing on himself and his intellectual pursuits. He embraced evolutionary theory, but he never felt convinced that science could explain creation — how life was spawned from nonlife. "It's very hard to understand how complex living organisms can be formed out of elemental, chemical molecules," he said. One night he woke suddenly and sat bolt upright in his bed. Hours later, he learned his grandfather had died. The experience shook him. Cynthia brought him to a service at her church, and he opened himself more fully to the spiritual realm. He saw the hand of God in many things, including his own path in life. "The definition of grace is 'unmerited favor,'" he said. "I feel like there's just a number of things in my life that have demonstrated unmerited favor."

As they were finishing medical school, the newlyweds struggled to find medical residencies in the same city. Cynthia, who was training to become a psychiatrist, suggested she'd try to land one at Meharry, a historically Black medical college in Nashville, while Graham applied to Vanderbilt University. Graham began his residency there in 1979 and only later learned that Tennessee had overturned its laws forbidding interracial marriage the year before he started. "If I had known that I may not have gone there at all," he said. That was grace, he believed. It was at Vanderbilt, after all, where Graham first dipped his toe into the RSV enigma.

During his quest to be the best doctor he could be, Graham became adept in photography, which he used to catalog all manner of medical ailments, filling up seven fat binders with slides. Anytime he came upon something unusual or significant about one of his patients, he took the time to publish it as a case study. One of his supervisors at Vanderbilt, William Schaffner, saw something special in the young doctor, and the moment Graham finished his residency, he was hired as a faculty member. Graham then did something that was, at that time, outlandish: he began working toward a PhD in microbiology at the same university. One of Vanderbilt's administrators learned of it and investigated the matter. Schaffner had to cover for him. Every time the administrator contacted him, he took a long time to respond. "I eventually wore him down," Schaffner said.

Graham's ravenous curiosity brought him to the topic of RSV. A measles relative, RSV is harmless in healthy adults, who may catch various strains of it multiple times in their lives without consequence. Among newborns, the elderly, and people with weakened immune systems, however, it causes up to 250,000 deaths each year around the world. There is no treatment for it, and in the 1980s, no one was willing to risk making another vaccine. The sickest RSV-infected infants, mostly preemies, were disproportionately Black and poor. Infectious diseases are not equal-opportunity killers. They often strike those members of the population with the fewest resources and the fewest advocates. Graham wasn't a pediatrician, but he was enmeshed in Nashville's Black community, attending the Mount Pisgah United Methodist Church, which was pastored by Cynthia's uncle. Graham knew that most of the children who had been enrolled in the RSV vaccine trial in the 1960s had also come from Black families.

While working on his dissertation, Graham collaborated with an NIH researcher named Brian Murphy who had reexamined blood serum stored from that tragic RSV clinical trial. One of Murphy's discoveries was that vaccinated infants had made a lot of useless antibodies — five ineffective antibodies for every useful one that neutralized the respiratory syncytial virus. Those extra antibodies didn't prevent cells from becoming infected, and they may have grabbed onto each other, forming clumps in the lungs that exacerbated the illness.

Between his hospital rounds, Graham studied this problem by recreating it in mice. He soon discovered that when the antibody response went off the rails, the rest of the immune system malfunctioned as well. You might say the immune system was confused about the size of the threat it was facing. A healthy immune response to a virus, as we've seen, involves, among other things, sending in killer T cells to destroy infected cells. But for larger interlopers that, unlike viruses, remain outside of cells — like bacteria, pollens, or parasitic worms — the body produces mucus, inflammation, and fighting cells known as eosinophils. This is the type of response we associate with allergies. That response cuts back on the killer T cells and generally makes it more difficult for the body to clear viruses from small airways. Graham saw that the vaccinated mice were essentially developing an allergic reaction against the clumps of antibodies rather than against the virus itself.

Over the next twenty years, Graham moved on to the NIH and sketched out a pathway that researchers could follow to ensure that a future RSV vaccine would not make the disease worse. The RSV vaccine had been made from a killed virus, but the chemical used to kill the virus had also deformed the virus's structure. If the body used a mangled virus as the template for its antibody response, those antibodies were going to be subpar. Instead of using the whole virus in an RSV vaccine, Graham concluded, it would be better to produce a vaccine based on a specific protein on the surface of RSV, called a fusion protein. Like the coronavirus spike, the fusion protein is the tentacle that tethers the virus to lung cells. On the surface of the virus, the protein is cocked and ready in what is called the prefusion state. After the virus has touched down on a cell, the protein snaps into a postfusion conformation to gain entry. Producing antibodies to the postfusion protein was like putting on body armor after you had already been shot. The problem was that nobody knew how to create a vaccine with the virus in the prefusion state. Nobody even knew what that prefusion state looked like.

And nobody wanted to help Graham find out.

When Jason McLellan showed up at the VRC as a postdoctoral researcher in 2008, he was planning to work on HIV. It was only through a happy accident that he landed on Barney Graham's floor. McLellan was supposed to be working with a guy upstairs named Peter Kwong, but there wasn't any office space left. McLellan was a humble fellow who had grown up on Detroit's notorious Eight Mile Road ("the slightly nicer side") and followed an unlikely path to the highest ranks of science. His specialty, structural biology, was becoming integral to the field of vaccine design. Compared to small, simple molecules in a tablet of Tylenol or an antibiotic capsule, vaccines were strange concoctions consisting of large, ungainly molecules that even their makers were at a loss to characterize or explain. Whenever McLellan's projects on HIV weren't working out, he'd come over to talk to Graham, because Graham was the kind of guy who everyone wanted to talk to. "He's one of the nicest human beings I've met, with no hyperbole," McLellan said. Ditto what Graham thought of McLellan. Every time the two men talked, Graham, a guy who knew more about mice than molecules, ever so gently insinuated that all the newfangled techniques McLellan was trying out with HIV might work in the hunt for an RSV vaccine.

Graham explained to McLellan how important it was to design an RSV vaccine with a less trigger-happy fusion protein. McLellan tried a few different techniques but nothing worked. Years went by. Then one night he was watching television and simultaneously trawling through Google's patent database. He found a patent from a company that had discovered two novel antibodies that could prevent RSV from infecting cells. Strangely, neither of the antibodies clung to the fusion proteins the company made in the lab. How were they stopping the virus? McLellan had a hunch that the company had found antibodies specific to just the short-lived prefusion form of the protein. He set up an experiment that produced both the antibody and the fusion protein simultaneously and found that the antibody was able to grip the protein in just the right way to lock it in its prefusion state.

Using a technique called X-ray crystallography, Graham and McLellan produced three-dimensional images of the antibody-bound protein, and the difference between the prefusion and postfusion structures was obvious. Prefusion, the protein was a squat triangle. Postfusion, it was like the Eiffel Tower, 50 percent taller. When the protein changed shape, it exposed different nooks and crannies, known as epitopes. This explained why so many of the antibodies people produced to the postfusion form were not neutralizing the virus. McLellan, Graham, and Peter Kwong (the guy upstairs) soon developed a technique to alter a couple of amino acids in their fusion proteins, making the structure stiff enough to be permanently locked in the prefusion state. This stabilized protein could provide a template for the body to produce neutralizing antibodies to RSV. In other words, a vaccine. Graham tested it in mice. The stabilized prefusion protein produced ten times as many neutralizing antibodies as the postfusion protein.

Graham was trying to crack one problem, to break down the barrier that prevented vaccine makers from developing a successful RSV vaccine. But as is often the case with the winding path of science, one discovery led to another. As Graham and McLellan were finishing up their RSV research, they heard about a new disease in Saudi Arabia. In June 2012, a man with acute pneumonia was admitted to a hospital there and had to be put on a ventilator. The standard virus tests came back negative. When the man died eleven days later, doctors sent sputum samples to a lab in the Netherlands, which identified the culprit as a coronavirus. They christened the disease Middle East respiratory syndrome, or MERS, and the virus that caused it seemed to have spread from bats to camels and then to humans. Ten months later, that same disease ripped through the Al Ahsa Hospital in the city of Al-Mubarraz, near the Persian Gulf. There were twenty-three cases and fifteen deaths. By August 2013, seventy-four people had been infected by the virus in Saudi Arabia. The Saudis restricted attendance at that year's hajj, the annual pilgrimage to the holy city. They asked elderly Muslims and those with chronic conditions to stay home.

Hadi Yassine, a Lebanese-American postdoc in Graham's lab, already had tickets for hajj in October. During his trip, one of his companions had become so sick he'd had to be hospitalized, and Yassine himself developed a bad cough on the flight home. When Graham heard that both Yassine and a companion were ill, he grew concerned that they had contracted MERS. A lab test revealed that Yassine didn't have MERS after all, but he did have another coronavirus, HKU1. For most people, HKU1 was as mild as a cold, though Yassine seemed to have gotten a bad case of it.

Graham was relieved by the news, but the scare added to the sense that he and McLellan should set their sights on developing an experimental vaccine for the MERS virus. The world had seen two coronavirus outbreaks in ten years. "It's going to happen again," Graham told McLellan. "We don't have any information on coronavirus structures, so this is something we really need to solve." As with the RSV fusion protein, the MERS virus spike spontaneously flipped from a prefusion state to a postfusion one shortly after it emerged from a cell. With the MERS virus, however, the team couldn't find the right antibody to clamp it in place. They were stuck.

Graham suggested they look at HKU1. They soon discovered that the prefusion structure of the HKU1 spike was more stable than the MERS virus's spike. They had seen the structure of the RSV protein using X-ray crystallography, but that process didn't work well for large, floppy proteins like the coronavirus spike, so they turned to Andrew Ward at the Scripps Research Institute in La Jolla, California. He flash-froze the spike protein, irradiated it with low-energy electrons to produce two-dimensional images, and then used thousands of these images to construct a three-dimensional model of that prefusion structure — the first-ever of a human coronavirus.

That mild virus, HKU1, became their prototype pathogen, offering them a way to crack the spike mechanism in more dangerous coronaviruses. In particular, they could pinpoint where on the original amino acid sequence of the HKU1 spike the flexible hinge region was located.

They could line the HKU1 hinge region up with the sequence for the MERS spike and begin to understand what it was about the individual amino acids that made the MERS spike so trigger-happy.

McLellan's postdoc Nianshuang Wang started modifying the spike gene to test ways to stiffen the MERS hinge and prevent it from thrusting into cells. Of the twenty amino acids used to make human proteins, proline was well known to provide the greatest stiffness to a molecular structure. Wang designed over a hundred variant sequences, inserting one proline in slightly different places. When he grew each of these modified spikes in the lab's cell cultures, he could measure, more or less, how many of them were locked in the prefusion state. With one of his variants, the number of stiffened spikes increased by a factor of ten. When Wang added two prolines to that same spot, he saw fifty times as many. Wang submitted the findings to *Nature,* one of the world's most prestigious scientific journals, but was swiftly turned down. The journal didn't think it was a significant finding. Over six months, Wang racked up two more rejections. "I was very depressed," Wang said.

Kizzmekia Corbett arrived in Graham's lab to begin her postdoc at just the right moment. At the NIH, she tested whether the S-2P design — the spike stabilized with two prolines — could work as a vaccine to protect against MERS in mice. It did. Wang incorporated Corbett's new data in his paper, and, after a couple more rejections, it was finally published in 2017 in the *Proceedings of the National Academy of Sciences.* The team had discovered what they believed to be a design for a vaccine that would target many, if not all, coronaviruses.

But they had yet to prove it worked in a real outbreak.

6

OLD SCHOOL, NEW SCHOOL

Before dawn on Monday, February 3, Gregory Glenn drove from his family vineyard in rural Maryland to Novavax's headquarters in Gaithersburg on Firstfield Road, a quiet street lined with two- and three-story buildings that had served as launching pads for a number of biotech companies. As a matter of fact, Glenn's first vaccine company, Iomai, had been in a red-brick structure across the street from Novavax, which had turned the building into its own manufacturing suite. But as Glenn idled his car in the parking lot, he could see that Novavax's logo had been removed from the red-brick building. Novavax had recently sold the lease, the equipment inside, and even, in a sense, the employees themselves. In the company's rocky three-decade history, it had never succeeded in bringing a vaccine to market and now it no longer had the means to make clinical-grade vaccines for human testing. How long the company could survive in this state, Glenn wasn't certain.

Another car pulled into the lot. It was an executive from GenScript bringing the coronavirus spike sequence at last. Unlike Barney Graham's team, which had ordered smaller gene sequences and pieced them together themselves, Novavax was competing with multiple companies to get its spike sequences synthesized. Two weeks earlier, the Novavax order was ready to be loaded onto a FedEx plane in China. But then the country's flights were thrown into disarray. Exasperated, Glenn picked up the phone, called GenScript's executives, and persuaded them to make the spike sequences again — this time in the company's New Jersey factory. Late on the night of Sunday, February 2, an executive loaded a package into his own car and drove four hours south to meet Glenn's team for the crucial early-morning handoff.

Novavax's tribulations had failed to sap Glenn's enthusiasm for his work. An army-trained pediatrician in his sixties, Glenn became energized when a company discovery meeting was on his schedule. Those were the meetings that featured the most cutting-edge science in the early stages of vaccine development. He and Gale Smith, the lanky scientist who headed the preclinical work, would file into a conference room on the second floor. The lights would darken. The projector would flip on. And there they were, one after another, the graphs and tables and three-dimensional images of molecules that looked like a bunch of billiard balls all stuck together.

Going back to the old platform Smith had developed, Novavax was planning to grow its spike proteins inside cells plucked from caterpillar ovaries. Although the concept had proven itself over and over again, every new protein you made in those cells required time-consuming tweaks to the purification and manufacturing processes, a big difference from the plug-and-play approach of Moderna's mRNA vaccines.

Novavax's general process began by harnessing the power of another virus, one that infected insects rather than people. When live caterpillars are infected with this virus, called a baculovirus, they ooze out a white goop until they die and dry up on a branch like an old scab, hard as a stone. If you examine that dried goop under a microscope, it looks like a cleaved crystal. That goop, as a young Smith learned in the late 1970s, is made up of a single protein, called polyhedrin, after its prismatic shape. The protein crystallizes around the virus like a snowflake, protecting it until another caterpillar comes along and swallows it days or years later. The crystals melt away in the gut of the caterpillar and the virus begins to multiply.

What was incredible was how much of this slime the caterpillars produced during an infection, likely more than any other protein in their cells. They became machines for sending viruses out into the world. Smith found a way to modify the genes in these baculoviruses to get caterpillar cells to manufacture any protein he wanted, including, in 1987, the HIV protein used in VaxSyn, the HIV vaccine that had been in multiple clinical trials. Although that vaccine was a failure, Smith's caterpillar cells made possible Flublok, Sanofi Pasteur's supercharged seasonal flu vaccine, and Cervarix, the human papillomavirus (HPV) vaccine from GlaxoSmithKline.

In 2004, Smith landed at Novavax. Glenn arrived six years later. Together, the two men were on track to bring the world's first RSV vaccine to market. The team had already started human trials in 2013 when Barney Graham and Jason McLellan published their stabilized prefusion protein. Novavax had ironed out the process to make a postfusion vaccine with its caterpillar cells. If the company started over with a stabilized protein, Glenn and Smith knew they would be overtaken by competitors. Novavax pressed forward, running a trial in nearly twelve thousand older adults in the United States.

Graham and McLellan could see during this time that Glenn wasn't pleased about the prefusion breakthrough. "Their thing was like, 'Well, maybe prefusion is not so great,' and then it was pure spin and marketing," McLellan said. By 2017, Glenn had stopped calling Novavax's vaccine a postfusion protein and started calling it "prefusogenic." Graham said, "After the prefusion concept got so much momentum and publicity, they made up a new term." Pretty much no one else in the scientific community were persuaded by Novavax's wordsmithing.

Dubious word choices aside, the thing that mattered was the clinical trial, and in September 2016, Novavax announced it was a flop. But the company doubled down, running a second trial in pregnant women in hopes of protecting their children. That one, too, was unsuccessful. Tens of millions of dollars down the drain. Novavax closed its vaccine-manufacturing suite and shed 750 of its 800 employees, including two-thirds of the research staff. Over the course of 2019, the company's stock price dropped 90 percent. Novavax's holiday party consisted of pizza and Coke in the break room. "I have PTSD from that," said Glenn.

When the new coronavirus broke out in early January 2020, would-be vaccine makers around the world started turning to Barney Graham for advice on their vaccine designs. Uğur Şahin, the CEO of BioNTech — Moderna's mRNA competitor, based in Mainz, Germany — told Graham he was planning to test several options at once, including Graham's stabilized spike, and he would compare the versions head to head in mice.

Dan Barouch, a Harvard researcher with ties to Janssen, Johnson and Johnson's pharmaceutical division, was planning to do the same thing in monkeys using an adenovirus as a vector, like the Oxford group was. As for the Oxford group — that brash, standoffish bunch — they were already charging down their own path: Sarah Gilbert was using the natural, unstabilized sequence in her vaccine. Who knew? Maybe stabilization wasn't all it was cracked up to be. Just because it seemed critical with RSV and the MERS virus didn't guarantee it would matter all that much with the new coronavirus. The jury was still out.

For Novavax, the stakes couldn't have been higher. A coronavirus vaccine was the thing, Glenn knew, that was going to either demonstrate the enduring value of protein-based vaccines or push them closer to obsolescence. The one thing Glenn was certain of was that the company shouldn't take any chances. Novavax would stabilize its spike proteins using Graham's method. Glenn evidently swallowed his pride and reached out to Graham, asking if there was a way that they might work together. Graham pointed them in the direction of his and his team's published papers and patents. Novavax was welcome to license the technology for its vaccine from the government.

Glenn and Smith would have to assemble their own collaborators from the academic world, a team that would need to compete with the might of the NIH and its big bet on mRNA technology. They would soon lock in their own structure guy — the microscopist Andrew Ward, who had worked with Graham and McLellan on the MERS project — and they also reached out to a researcher in Oklahoma who could test Novavax's vaccine in primates.

The first collaborator they brought on, however, was Matt Frieman, a virologist at the University of Maryland who had access to his university's biosafety level 3 lab, where he could work with the new coronavirus. Weeks earlier, Frieman had been on the verge of shutting down his coronavirus program. Almost all of his grant proposals had been rejected over the previous seven years. "I didn't have enough money to run the lab," he said.

Once the outbreak was confirmed to be caused by a coronavirus, he was eager to be a part of anything that developed. He pinged a contact at the CDC to ask if there was any way "to get involved." A few days later, he reached out to a biotech company. "Any interest in working on Wuhan coronavirus?" Frieman wrote. No dice.

Then, on Tuesday, January 21, Frieman received an e-mail from Smith at Novavax. "It seems coronaviruses are back in the news," Smith wrote. He asked if Frieman was game. Frieman pondered the prospect for no more than a few minutes. He knew that he would have only a small role in the vaccine-development process, a pit stop on the way to human clinical trials, but without the data he could provide, neither the company nor the regulators would have the confidence to move forward. He told Smith he was in. The first thing Frieman needed to do was obtain live virus. As Yong-Zhen Zhang had demonstrated eleven days earlier, genome sequencing allowed the scientific details of new pathogens to zip instantly around the globe as digital bits, which meant that you could manufacture a vaccine candidate within days of receiving the sequence. Biological samples such as viruses, however, typically had to be grown inside laboratories and shared the old-fashioned way, through courier services like FedEx and DHL. Live viruses were needed to evaluate how well the antibodies prevented infection and to run the critical challenge trials, where vaccinated mice or monkeys were deliberately exposed to the virus to assess the vaccine's effectiveness. Moving those viruses across international borders required getting through red tape, and, as in the case of China, a country might decline to share a sample at all. But even after the virus had found its own way into the United States, researchers were still hunting for samples.

One of the people Frieman texted was friend and former colleague Natalie Thornburg, an immunologist at the respiratory viruses branch of the CDC. After the first U.S. patient, a Washington State resident, was diagnosed on January 20, the virus-containing vials were frozen and shipped to the CDC's headquarters in Atlanta. Two of Thornburg's senior scientists then divided up the virus sample into ninety-six wells on a tissue-culture plate. They added Vero cells, which the viruses would infect in order to replicate. Named for the Esperanto words *verda reno* ("green kidney"), these unusual cells were first taken from the kidneys of an African green monkey and have an abnormal number of chromosomes. Most cells age each time they make a copy of themselves, but these cells are "immortal," meaning they can be kept in laboratory cultures indefinitely. Thornburg's team kept her infected-cell cultures at body temperature, sustaining them on cow blood and fending off bacterial trespassers with antibiotics.

Each day, her team examined the Vero cells for evidence that coronaviruses had gained entry. Under a microscope, healthy Vero cells look like little kayaks, elongated and pointed at either end, with a round structure in the center, the nucleus. Once infected, the cells bubble out and become foamy-looking, and their membranes eventually burst. The moment the cells start dying, Thornburg's team had to sequence the contents to ensure that a coronavirus, rather than some other random virus, was responsible. Then they scraped a single layer of cells from their culture plates and diluted it in order to grow more of the still-unnamed coronavirus. "Someone needs to name this darn thing so we know how to label our tubes," Thornburg tweeted in late January.

On January 30, Thornburg sent an e-mail to Frieman telling him he had made the "varsity team." She was sharing her virus stock with a facility at the NIH, which would start growing more of it for the research community, but she was also going to send some directly to Frieman and eight other labs. A week later she tweeted, "Lots of frustrations this week, BUT first round of viruses shipped out." Not long after, two small tubes of the virus arrived in Frieman's mailbox. Frieman stuck one of the tubes in a freezer and cultured the other one in the lab. If the virus mutated over time, he could always go back and start again with the other tube. After a few days, he was ready to test the virus in his mice.

One morning Frieman and his graduate student Rob Haupt zipped white PPE suits over light blue medical scrubs and clipped powered respirators around their waists like fanny packs. The accordion-like tubes from their respirators connected to the backs of their hoods and face shields. Frieman flipped the switch, and his hood grew taut as air blew across his face and out at the neckline. All he could hear was the whir of the motor. It was like scuba diving. The two men passed through the door into the BSL-3 lab. Shelves holding clear plastic boxes, each one labeled and filled with mice, lined the lab.

Frieman pulled the first mouse out of the box and injected it with ketamine, an anesthetic. Then he passed it on to Haupt, who tagged its

. ear with a numbered clip made of metal. Next, Frieman took a pipette filled with virus, squirted it into the mouse's pink nose, and placed the mouse back in its cage. It would wake up a few minutes later. Each day, Haupt or Frieman placed each mouse on a tiny scale and recorded its weight. The mice were all holding steady at twenty grams or so. They didn't seem to be getting sick. When Frieman dissected one of the mice after a few days, he saw perfectly healthy pink lung tissue. Other researchers were finding the same thing. The first SARS virus had been able to infect mice, but the spike of the new coronavirus had a change in a single amino acid in a key region. That made the spike different enough that it couldn't latch onto the mouse version of the human ACE2 receptor. Frieman needed a plan B. He reached out to a lab in Wisconsin that was working on a genetically modified cold virus that, like a scientific magic trick, would give his mice human ACE2 receptors, making them susceptible to this new coronavirus.

It would take a couple of weeks before that happened, but, like the NIH team and other researchers collaborating with vaccine makers, Frieman had tools to develop and tests to run before Novavax's candidate entered phase 1 trials. The company was advancing its protein-based vaccine as fast as any company ever had, but would that be fast enough in the face of a competitor like Moderna?

Perhaps it should be no surprise that most Americans still viewed the coronavirus as someone else's problem. It was like hearing about a mudslide in South America or a famine in sub-Saharan Africa. It was another tragic, isolated event that barely rose above the noise of daily routines — the children that need to be taken to school, the presentation that the boss was demanding.

Even members of the public health community were expressing the sentiment that media reports about the coronavirus were overblown. It was more disaster porn, a head-turning distraction from the real challenges in global health. On February 2, China released the latest statistics on the outbreak, tallying 361 deaths and 17,205 confirmed cases, which one user on Twitter called "grim." Joel Selanikio, a former CDC epidemiologist, fired back this retort: "Struggling to understand 'grim' in the context of >15k flu deaths this season in U.S. alone, and very likely more in China," he wrote. "What words do we reserve for the 250k traffic deaths, or 1.5 million tobacco-related deaths, annually in China?"

Kizzmekia Corbett, who had come to work before dawn on the morning of February 3, responded to Selanikio's ill-considered tweet. "My lord . . . take the spotlight from the flu biologists and they are throwing tantrums," she tweeted back. "It's almost like the 'all lives matters' crowd but for virologists. Yes, beloved . . . all respiratory viruses matter thus coronaviruses matter."

Corbett and Graham knew that others in the scientific community were expressing more specific doubts about the prospect of a coronavirus vaccine. The coronavirus might evolve to evade immunity, for instance, or the neutralizing antibodies to the spike protein might not be sufficient to prevent full-blown infection. After all, about 15 percent of colds are caused by coronaviruses, yet there's no vaccine against them. People can be repeatedly infected with coronaviruses causing the common cold, though it still wasn't clear whether this was because immunity fades or the viruses change. This isn't the case with polio or measles, where a natural infection provides sterilizing immunity for decades.

The way the human body is supposed to fight a familiar pathogen goes like this: Once a virus gains entry, the body produces special cells called plasma cells. These are antibody factories that pump out up to two thousand disease-fighting molecules every second. This process is metabolically expensive, so after an infection wanes, plasma cells taper off, as do circulating antibodies. Plasma cells generally live for a few days or months at most. Some of those cells, however, will remain in the bone marrow for several years, secreting low levels of protective antibodies into the blood just in case. Another population of immune cells, known as memory B cells, then move into places like the tonsils and lymph nodes, where they wait, like night watchmen. Rather than secreting neutralizing antibodies, the memory B cells remain attached to the outside of cell membranes like little probes. If those old memory cells detect that virus again, they rev up and make another round of plasma cells.

The success of vaccines hinges on those memory cells. But with coronaviruses, some feared that our bodies had amnesia. In one study, the memory cells of people who had been infected with the first SARS virus, for instance, failed to reactivate when they were exposed to the same virus six years later. If true, this would mean the body had to start fresh with the process of selecting the best antibodies from the galaxy's worth of antibodies in its arsenal. A coronavirus would then always have a chance of slipping past antibodies, the body's first line of defense, and entering its cells. The only way to stop it, at that point, was for the body to blow up its own cells via the more complicated, more fraught killer T cell defense. Naysayers argued that a quick-and-dirty coronavirus vaccine was a fever dream. A vaccine might be only partly effective and immunity would vanish within months, and we'd all be back to square one.

Graham felt in his gut that the prospect of coronavirus immunity wasn't far-fetched. A vaccine might well be better than natural immunity. One of the early worries about coronaviruses, as with RSV, was that the human body wasn't good at generating a neutralizing antibody response to a natural infection. The susceptible region of the spike, the lockpick, was exposed for only a short time before the spike flipped into that postfusion conformation. This seemed to be a tool to confuse the body's defenses. But Graham believed that when the spike was successfully fixed in place in its prefusion conformation, the human body could learn how to beat back the virus for a good long while.

The proof was coming—some of it, at least. On Tuesday, February 4, Corbett repeatedly refreshed the tracking number on a package from Moderna heading her way—the first batch of its vaccine, called mRNA-1273. When it arrived, however, the courier wouldn't even set it on the loading dock until Corbett went downstairs and showed her ID. Next, she rounded up several colleagues, and they hurried into the basement to the animal-care room, where the mice were maintained by a staff of veterinarians and aides. Corbett and her team took blood samples and then injected three groups of ten with different doses of the vaccine. A fourth group received a saltwater placebo. A pro like Corbett could vaccinate one mouse every minute. The goal, at this stage, was to show that the vaccine could stimulate antibody production. That step had to occur in order for them to begin the human safety studies. After that, the team would continue working with mice and monkeys to develop stronger proof of efficacy before expanding the human studies and putting more volunteers and money at risk.

Although it seems obvious today that drugs and vaccines should be tested in animals before they're tried in people, that's a surprisingly recent development. It wasn't until 1938 that the U.S. Food and Drug Administration required companies to submit safety data before putting their products on the market. But even then, the agency wasn't explicit about what tests companies needed to run. And there were no requirements whatsoever for launching a human clinical trial. During a 1958 congressional hearing, a doctor named Louis Lasagna bemoaned the sorry state of the agency's regulations. "It is reprehensible for man to be the first experimental animal on which toxicity tests are done, simply because bypassing toxicity tests in laboratory animals saves time and money," he said.

Five years later, the FDA finally updated its regulations, creating a formal path for companies and organizations to test a new product in people. Before beginning a phase 1 clinical trial, the research sponsor had to file an Investigational New Drug application, or an IND, providing basic animal toxicity data showing that its drug or vaccine was likely to be safe in humans, at least in the short term. The IND also includes information about the company's manufacturing process, the protocol for the first-in-human safety study, and why the company thinks the product has a reasonable chance of working.

Moderna had farmed out the animal-toxicology tests to contract laboratories, while the NIH handled the immunology studies. Around noon on February 19, the three students working for Corbett began testing the vaccinated mice for spike antibodies. They placed blood samples into centrifuges to separate the dense red blood cells from the lighter, antibody-containing serum. Then they prepared plastic plates containing dozens of wells with the spike protein stuck to their surfaces. They added serum from each of the mice into these wells. After incubating the plates at body temperature for about an hour, they rinsed them out. If the serum had contained spike antibodies, those antibodies were now attached to the spike proteins that were still adhering to the plate.

In order to see if that had happened, they added to the plate a specially made detection antibody. These clamp onto other antibodies and carry a color-changing enzyme. The team members washed the plate a second time, flushing out any unbound antibodies. Finally, they added a chemical that triggered the enzyme to change color. The brighter the yellow, the more spike-binding antibodies the mice had developed. This test, which is standard in immunology laboratories, is known as an enzyme-linked immunosorbent assay, or an ELISA. One of the students, Cynthia Ziwawo, recalled seeing the first plates turn yellow almost instantaneously. The vaccine was causing the mice to make antibodies to the spike protein.

Whether or not those antibodies could neutralize a virus, no one could say. Determining that required placing antibody-containing serum in a dish with live cells and seeing if it prevented a viral infection. They were going to have to run hundreds of such neutralization assays in the coming months for all their different experiments with mice, monkeys, and, eventually, people.

Rather than performing all those tests with a live coronavirus inside cumbersome biosafety level 3 conditions, Corbett and her team would do something that sounds ridiculously complex but has become rather workaday in the world of virology: they would construct what was known as a pseudovirus. This consisted of a set of genetic instructions packaged inside a protein coat covered in coronavirus spikes. Like the real coronavirus, the pseudovirus docked on a cell with one of its spikes and then injected its genes, but those genes didn't tell the cell to make copies of the pseudovirus. Instead, they told the cell to produce luciferase, the enzyme that makes jellyfish glow green. When the researchers later split the cells open and added a chemical fuel, the infected cells would light up. If the cells didn't light up, Corbett would know that the pseudovirus hadn't entered the cells, meaning their mice had produced a sufficient amount of the coveted "neuts," the neutralizing antibodies, to prevent infection.

The NIH team, however, wasn't going to wait the month or so for the pseudovirus-neutralization assays to be completed before asking the FDA for permission to start the human trial. The FDA needed only to be convinced of the promise of the vaccine and have data showing that it wasn't acutely toxic, data that Moderna provided from its own studies. Graham waited around the office to get the first peek at the ELISA results late that evening. During a call with Moderna the next morning, February 20, he told the team the good news: all systems were go.

Moderna filed its IND application with the FDA that day. Four days after that, a truck pulled out before dawn from the company's factory in Norwood, Massachusetts, and headed southwest, passing through New York City and Philadelphia on its way to the NIH with the first human-grade batch of mRNA-1273. Tucked inside the precious cargo was a GPS tracker and a data-logger recording the temperature every couple of minutes. Unlike Novavax's vaccines, which simply needed to be refrigerated, the mRNA vaccine was so fragile that it would start to decay if it wasn't kept frozen, especially during transport. If at any point along the journey the temperature inside rose above Moderna's specifications, the company would have to discard these doses at great expense and start over again.

Stéphane Bancel was on tenterhooks for the entire eight-hour journey. And for good reason: Back in 2015, when the company was preparing for its first-ever vaccine trial, for an avian flu vaccine, the box containing Moderna's vaccines sat cooking atop the hottest part of the truck. By the time the vaccine arrived in Miami, it was toast. Moderna made the vaccine again and shipped it a second time. It survived the trip, but the freezers at the clinical trial site failed to keep it cold enough. The Miami fiasco had cost the young biotech company several hundred thousand dollars.

This time, there were no mistakes. At NIH that evening, the vaccine was pulled off the back of the truck and moved to a freezer bank. As the NIH team awaited the green light from the FDA for the first human study, Corbett and Graham pressed on with the animal work, which was far from complete.

7

OCEANIC OUTBREAK

One month into the coronavirus outbreak, most Americans had yet to be inconvenienced. In late January, a cruise ship called the *Diamond Princess* had sailed out of Japan and made port in Vietnam, Taiwan, and Hong Kong. The vessel rose thirteen stories above the waterline and stretched the length of three football fields. It was a great, glistening, white behemoth — a city on the sea.

There were more than thirty-seven hundred people on board, including Arnold and Jeanie Hopland from Tennessee. The Hoplands had spotted a last-minute deal and booked themselves a cabin with a porthole on the lowest deck. They were in their seventies and it was their first time in the Far East. It was nippy out on the Pacific, and Arnold Hopland liked to soak in the Jacuzzi on the sundeck until his face turned pink. In the evenings, he'd skip the shows and hang out at the piano bar drinking virgin cocktails and trading corny one-liners with the musician.*

When the vessel docked for its second-to-last stop in Okinawa on Saturday, February 1, the Hoplands were told they'd need to have their temperatures checked at port before they could go on a tour. They waited in a long line as Japanese workers reviewed their documents and took forehead temperature readings. By the time the Hoplands made it through, it was too late for their excursion. Instead, they ambled around the city on their own for a couple of hours and then returned to the ship.

---

* "Did you hear about the new corduroy pillows? They're making headlines."

They would be disembarking in Yokohama early the next morning, so they packed their bags and set them outside their door.

After dinner, the couple was playing a board game when an announcement came over the loudspeaker: Everyone's departure would be delayed for twenty-four hours while Japanese authorities inspected the ship. Hopland and his wife didn't think too much of it. The following day, they were given the news: A *Diamond Princess* passenger who had disembarked in Hong Kong on January 25 had tested positive for the coronavirus. Now, eleven days later, ten people on board were also infected. All passengers would be quarantined in their cabins for the next fourteen days.

Back in the United States, health secretary Alex Azar needed money. The repatriation flights from Wuhan and the airport screening efforts were churning through the $150 million he had in a special HHS fund for rapidly responding to infectious diseases. The country's health-care system had on hand only two weeks' worth of disposable N95 respirator masks, which were needed to protect hospital workers from infection. Most masks were manufactured abroad, as Bob Kadlec had warned officials, and many countries were already locking down exports. In the event a national outbreak became a reality, the United States would burn through the twelve million masks in the already-depleted national stockpile within weeks. Dr. Bob had negotiated a deal with 3M for five hundred thousand masks, but ASPR had no cash left to buy them.

President Trump's State of the Union address would be taking place in the House Chamber of the Capitol Building at nine p.m. on Tuesday, February 4. That evening, as Azar waited inside an ornate retiring room preparing to proceed onto the floor with other members of Trump's cabinet, he cornered White House budget director Russell Vought. It was critical, he told Vought, that the White House submit a supplemental budget appropriation request to Congress. "The pressure is building," he said. Azar was due to give a closed-door briefing on Capitol Hill the next day and said that the administration couldn't let the Democrats control the narrative. "Why take the political hit?"

The bald, bespectacled Vought grimaced. He had always had a dim view of the Strategic National Stockpile and of the for-a-rainy-day requests coming in from Kadlec's shop. He told Azar he would consider asking for appropriations, but the amount had to be small. "It can't be a wish list," he said.

Azar's team drew up an internal budget figure based on the demands coming in from the CDC, ASPR, and NIH amounting to about ten billion dollars, as Kadlec had estimated. Azar knew the way the game was played: no matter the number approved by Vought's office, the free-spending Democrats in Congress would surely increase it. He cut the internal request in half and then cut it a little more for good measure. The next day, the health department submitted its request for $4.5 billion, including around a billion for ASPR, to Vought.

Azar sat through a grilling by Democratic representatives inside the Capitol. When Rosa DeLauro, a senior member of the appropriations committee, asked him about the administration's plans to request more funding, he remained tight-lipped, as per protocol. He said only that HHS would present a request when it needed supplemental funding.

The next day, Vought came back and told Azar that the sum he had requested was over the top. Vought said that in six days, on February 12, he would be testifying about Trump's 2021 budget request and that Azar would have to table the supplement talk until after that date. The coronavirus could wait.

On Monday, February 10, Michael Callahan arrived at the Humphrey Building for his first day of work. When one of Kadlec's executive assistants came to retrieve him, he held out his hand to introduce himself. Callahan just stared at it. Two weeks in a hot zone will do that to a person.

After departing China, he had stopped over in Singapore to meet colleagues dealing with the coronavirus there and then changed his ticket home so he would be landing at Dulles Airport near Washington rather than in Boston. Once inside Dr. Bob's office, Callahan replayed, in his amped-up, breathless mode of speech, the horrors that had befallen Wuhan's hospitals: patients in the hallways, health-care workers abandoning their posts, oxygen supplies running low.

"Holy shit," Kadlec said to his old friend.

That week, the World Health Organization gave the disease its official name: COVID-19, short for "coronavirus disease 2019." The virus was given a name as well: SARS-CoV-2. Kadlec knew that ASPR would soon be called to play a larger role in the nation's response to the growing threat. But when? How? Who would be making the call and on what basis? Would it be Alex Azar? The task force? The president? At this particular juncture, all eyes were on the CDC, and the agency's view was that it was all about containment, about keeping the virus outside the U.S. through airport screening and track-and-trace programs. "I defer to my experts" was Azar's line at task force meetings. He was adamant about always bringing Tony Fauci and Robert Redfield with him to speak to the press or to brief the president. He had faith in the ability of science to inform the decision-making process. But as the outbreak worsened, that faith would be challenged.

The CDC had tested 330 people at that stage, and just 12 had been positive, which sounded encouraging. But some officials began to wonder how many cases might be slipping under the radar. "If cases start occurring in communities, we will not get a second chance at keeping the confidence of the American people," Ken Cuccinelli, the acting deputy secretary of Homeland Security, warned in an e-mail he sent to Azar and other members of the task force over the weekend. "I propose that we allocate significant time to discuss shifting our operational posture to be more aggressive on the mitigation front."

There would soon be a collision between those in the wait-and-see camp and those in the we've-got-to-act camp. The Japanese were being pressured to break the quarantine on the *Diamond Princess* and release the passengers early. On Monday night, Kadlec e-mailed Japan's vice minister for health, Yasuhiro Suzuki, offering to deploy a medical team immediately. In the interest of preparedness, Kadlec and Suzuki had discussed a disease-outbreak exercise in advance of the planned Tokyo Olympics. "As we've talked for the past 18 months about such a scenario, this may not be the one we envisioned, but it is a perfect case," Kadlec wrote. "Our mutual efforts and planning have paid unexpected dividends."

By seven o'clock the next morning, Shinzo Abe, the prime minister of Japan, had approved the operation, but Kadlec couldn't rally the support he needed from others in the government. Another day went by, and a total of 135 passengers had now tested positive for the coronavirus. This didn't sound good to Kadlec, but on Wednesday he attended a congressional briefing concerning the cruise ship, and Anne Schuchat, the principal deputy director of the CDC, said that the Japanese were indicating that the situation was under control. The passengers were better off remaining in their cabins and under quarantine.

When she finished speaking, Phil Roe, a congressman from Tennessee, ripped into her, telling her that his constituent Arnold Hopland was trapped on the cruise ship at that very moment and had told him that things weren't under any kind of control. Hopland had called the ship "a petri dish." The virus was not being contained. Masks and gloves hadn't been handed out quickly enough, and crew members were falling ill. "We need to evacuate that ship," Roe said.

Kadlec was now determined to move forward with the operation. After the session, he approached Roe and asked if he could talk to Hopland directly. That evening at nine, Kadlec convened an emergency call, bringing in officials from the CDC and the State Department. They phoned Hopland, and he reported the dire situation on the ship.

The State Department and the CDC gave Kadlec approval for an evacuation as long as he agreed not to repatriate anyone who tested positive for the coronavirus. Infected passengers would have to be quarantined in Japanese hospitals. This kind of complex and risky operation had never been done in U.S. history, and there were many ways it could go wrong, but no one could accuse Kadlec of being risk-averse. After the call, he told Suzuki that they were moving forward.

Callahan was in Boston helping his staff prepare for the inevitable crush of patients when Kadlec called him and told him to get his ass on an airplane. Callahan threw his business suit and winter coat into a FedEx box and mailed them to his home in Colorado, then headed to Target and bought a hundred and fifty dollars' worth of cheap polyester clothes that he could throw away at the end of the operation. He flew from Boston's Logan Airport to LAX and met up with one of the Wolverines, James Lawler. Lawler had picked up twenty-one powered air-purifying respirators (PAPRs), which the two men split between their suitcases.

They flew to Tokyo, where they were joined by three of Kadlec's disaster medical-assistance teams, which included doctors, nurses, and emergency medical technicians. Over the course of forty-eight hours, the teams set up a dispatch center on the *Diamond Princess* and screened the four hundred or so American passengers, identifying those with mobility issues or chronic problems that would make them most at risk if they were infected.

After ten days trapped in their cabin, the Hoplands were eager to depart. They heard a knock on their door on the night of February 16. Jeanie put her coat and mask on and opened it up to see a Japanese man standing there. "You tested positive," he said. "You can't go." Jeanie would have to check into a hospital; Arnold was welcome to fly home. But of course, he had to have the bug too! "The test was five days prior, and I'd been living with her the whole time," he said. He decided to stick out his quarantine for another ten days on the ship and return home with Jeanie after she was released from the hospital.

Meanwhile, after other passengers had already piled into buses heading to the Tokyo airport for flights back to the United States, word came in that fourteen of them were also positive. It was already late. The airfield was deserted. It was raining sideways. Callahan was keen to inspect the layout of the plane, but he and Lawler were told they needed to cool their heels in a nearby building. The moment his minder looked away, however, Callahan booked it out to the jet and charged up the stairs to scope it out.

As the buses idled for the next four hours, Kadlec, back in DC, traded harsh words with his colleagues. The CDC and the State Department were both hesitant to move forward with the flights. Kadlec believed the doctor's oath to save the sick trumped the public health instinct to contain the virus. He pointed out that the State Department team had hung ten-foot-high sheets of plastic on the two planes to isolate any passengers that developed symptoms during the twelve-hour flight. "The culture I came from is leaving no American behind," he snarled.

Considering the multiday delays in receiving results, there were bound to be other infected passengers heading home. With the clock ticking for the aircrew and no viable alternatives for the passengers, the State Department sided with Kadlec. The CDC withdrew its support for the operation. The planes nevertheless departed. Callahan and Lawler spent most of the flight suited up in the passenger compartment, the motors of the air purifiers humming the entire time.

Once in the United States, the two doctors escorted their sick patients to the medical isolation unit Lawler managed in Nebraska. The evacuees were going to be quarantined on military bases. But the military changed its tune when at least sixteen individuals who had previously tested negative tested positive. Inside the Secretary's Operations Center, Kadlec's deputy Kevin Yeskey was up all night making calls to hospitals and state and local health departments, trying to find rooms. "Just do this for me once," he begged. "Take this guy." He managed to find space for up to fifty people in a shuttered mental-health facility in Costa Mesa, California, but then the city sued the feds, and the plan was abandoned. It was disheartening, but Kadlec's team struck deals and found alternatives to house the evacuees.

On Thursday, February 20, the *Washington Post* published a story about the confusion on the ground in Japan, implying that Kadlec was tearing down the fortress of defenses the CDC had supposedly erected to keep the coronavirus out of the country. It was Kadlec and his disease cowboys, in other words, who were the greatest threat to the country. President Trump read the article and called Azar in a huff. He wanted to fire whoever was responsible.

Azar hemmed and hawed. He could either send Kadlec out as a sacrificial lamb or admit that the CDC's containment strategy was becoming a farce. He eventually said that it was a tough call on the ground, but it was really the chargé d'affaires at the Tokyo embassy who had pulled the trigger on the decision. That was the State Department, not HHS. "Talk to Mike Pompeo," he said with a shrug, giving cover to his airman in the Humphrey Building.

Because CDC officials were unwilling to admit that they'd failed to keep the virus out of the country, the rest of the government's pandemic apparatus was effectively paralyzed. In departmental meetings, the CDC insisted there was no reason to move forward with school closures and social-distancing measures until there was proof—absolute proof—that the virus was spreading in communities. For the agency, that meant scientific evidence that the virus had been transmitted from one individual to another and from that person to a third in multiple locations inside the United States. The fundamental problem was that the CDC was not running nearly enough tests at its headquarters in Atlanta to have any hope of meeting its own high evidence bar, and the test kits the agency had sent out to state labs weren't working properly.

On the afternoon of Friday, February 21, Kadlec and Alex Azar headed to a task force meeting to run a tabletop exercise, gaming out how the outbreak might develop. They placed their cell phones in lockboxes outside the Situation Room, a sensitive compartmented information facility, or SCIF, in the basement of the West Wing. Seated around the oval table were the task force members from various agencies, including Transportation and Homeland Security, along with White House advisers and staff.

Over the next two hours, Kadlec broke down two scenarios, coded red and blue, respectively, on the handouts he gave to task force members. Under the red scenario, he envisioned the novel coronavirus exploding in small pockets like the first SARS outbreak had; these could be controlled through aggressive local public health responses. There would be little risk to medical-supply chains or critical infrastructure under this scenario. However, Kadlec told the task force, there were signs that the clusters were already expanding beyond the point of containment. This brought him to the blue scenario, in which a more insidious and less severe virus blanketed the whole country. Cases would double every five to seven days, as they had in Wuhan. Medical supplies would have to be rationed, and hospital capacity would be insufficient to handle all the patients. Reducing the spread would require "extensive community-mitigation measures" and a "national-level response" that his agency, ASPR, would lead.

Before the meeting adjourned, Azar brought up the topic of money again — the now-stagnant coronavirus supplemental appropriation. The following week, he said, he would be testifying before multiple House and Senate committees about the 2021 budget. The Democrats would eat him alive. Mick Mulvaney, Trump's chief of staff, looked around the table. There was no way Azar would be testifying without a supplemental being sent to the Hill first, he said. "Make it happen."

The next morning, a Saturday, as Azar was having breakfast at the

Daily Grill inside the Hyatt Regency in Bethesda, he called Russ Vought of the White House budget office to check in on the appropriation's status. The $4.5 billion that Azar had requested, Vought said, would have to be whittled down to $1.25 billion in new funding, along with a transfer of $535 million that had been set aside for Ebola. Kadlec would be granted just $515 million out of that pot, enough to buy some N95 face masks and gloves and little else. Any major investments in vaccines and therapeutics would have to wait until September 2020, when the next fiscal year began.

8

VERY MUCH UNDER
CONTROL

When Michael Callahan returned from Japan, he was not his best self. He threw his bag into Bob Kadlec's office and headed to the charmless conference room next door. Seated at the long table with more than a dozen staff was Rick Bright, there to lead a meeting of the recently established coronavirus medical-countermeasures task force.

Callahan and Bright went way back. Bright had worked on diagnostics at the CDC and on vaccines at Novavax and PATH, a global-health organization. When Callahan was at the Defense Department, he needed a vaccine guy during the 2009 swine flu outbreak and he hired Bright as a consultant. The next year, Callahan recommended him for a position at BARDA, ASPR's drug-development division. Unlike Bob Kadlec, a political appointee, Bright was one of the nation's two million career government employees who enter civil service through a competitive hiring process and rise through the ranks.

He was now the director of BARDA, taking early-stage research coming out of private companies, universities, and even the NIH and helping to get it onto the market. Although the NIH could enter into collaborations and give research grants worth tens of millions of dollars to companies, only BARDA had the statutory authority to sign hundred-million-dollar development-and-acquisition deals for commercial products. With a raging coronavirus outbreak, antiviral drugs represented one means to save the sick and keep patients from overloading hospitals.

While that task was certainly not easy, the prospect of manufacturing and stockpiling tens of thousands or even several million doses to treat the infected was far more plausible in the near term than developing and producing the hundreds of millions of vaccine doses needed to immunize the healthy population.

Bright had close-cropped gray hair, a smooth, boyish face, and a nasally twang to his voice. He piped up almost as soon as Callahan stepped into the conference room. "Well, if Dr. Callahan would answer his e-mail and be responsible for his clinical responsibilities to the ASPR," he said.

"What are you talking about?" Callahan glared at him. While Callahan was in Japan, Bright had been pressing him to get an unapproved antiviral drug to the American patients who had been taken off the *Diamond Princess* and placed in Japanese hospitals. That was rich, Callahan thought. This laboratory scientist — or "mouse doctor," in Callahan's words — sitting in an office in DC was telling the battle-tested disaster doc in the field what to do. The drug Bright was pushing was called remdesivir and it was made by Gilead, a biotech company notorious for the eye-popping prices it charged. Its hepatitis C drug cost $1,100 per pill.

In Callahan's view, remdesivir was a drug looking for a disease — or at least a pocketbook. During the 2014 Ebola outbreak in Liberia, Callahan had seen it fail in a major way. It had to be administered intravenously and required multiple injections and close monitoring — a big ask for a doctor working inside a dimly lit tent, sweating in the tropical heat in a biosafety suit. Even worse, local doctors noticed it showed signs of causing liver toxicity, and Ebola patients were often already on the verge of liver failure. For the new coronavirus, Callahan's Chinese contacts had told him that remdesivir was no miracle drug, and Callahan wasn't about to recommend it for the *Diamond Princess* patients. "They were sick as dead," he said, using one of his idiosyncratic expressions. "I'm not going to give them a known hepatotoxic agent."

But Bright wouldn't let it go. "There are other clinical experts who disagree with you, Dr. Callahan," he chided. Callahan became livid. He laid into Bright in front of everyone, belittling his professional qualifications. Kadlec's chief of staff, Bryan Shuy, told the two of them to knock it off. Afterward, Bright sent Callahan an e-mail assuring him that they were on the same team. Callahan regretted his explosion, but whatever team he was on, it was not Bright's. "I just put him in his place," Callahan said. "It was not very professional on my part."

One of the most pressing questions at this juncture was whether asymptomatic people were spreading the virus. It all depended on how much virus was in the nose and upper airways, where it was readily transmitted through coughing and sneezing. When James Lawler asked the CDC for permission to test asymptomatic evacuees held in quarantine in Nebraska, the CDC's Robert Redfield and Anne Schuchat refused on ethical grounds; they said those individuals might feel like they had no choice. *Are you kidding me?* Lawler thought.

The numbers coming from the *Diamond Princess* confirmed Lawler's worst fears. Of the four hundred Americans on board, fifty-eight were confirmed to have infections. Of those, twelve were asymptomatic and forty-six were symptomatic. About a quarter of the infected patients required hospitalization, and 2 percent needed intensive care. Bob Kadlec was shocked to hear these numbers, having been told again and again by the CDC that the situation was under control.

On February 24, Callahan and Lawler traveled to Atlanta to speak to the CDC's leaders. As government outsiders, they were hoping to air their grievances over how the repatriation and evacuation efforts had played out and broker a peace deal with the career officials, who they suspected were taking anonymous swipes at the ASPR in the press. At CDC headquarters, the two men headed up to the twelfth floor and entered the director's office, where Redfield and a number of top officials, including Schuchat and Marty Cetron, the CDC's quarantine director, were gathered.

Callahan reminded Redfield and Cetron that he had begun his career at the CDC. But he was, he said, dismayed by the sloppiness of the agency's infection-control practices in the field. Lawler then made the case that it was no longer enough to test individuals with symptoms. Redfield didn't say much. Afterward, Lawler and Callahan took the elevator down with a CDC scientist who had been in the meeting. The scientist said, "Well, if we find there is community transmission —"

"Come on," Lawler said. "We *know* there's community transmission."

"I know, but we haven't proven it yet" came the reply. Lawler and Callahan exchanged looks of disbelief.

At the time, there were fewer than sixty officially confirmed coronavirus cases in the United States, but no one knew what the true number was. The test the CDC had sent out to state health laboratories was still not working. One of its components had been contaminated at CDC headquarters. By the time Callahan and Lawler returned from their trip to Atlanta, the CDC was heading for a reckoning, and so was the man who put so much stock in its scientists: Secretary Alex Azar.

Over the next two days, a comedy of errors played out that was funny to no one. President Trump, who was touring India with First Lady Melania Trump, reiterated to the press that the virus was "very much under control." On February 24, the task force agreed that when the president returned from India, Kadlec and Azar would deliver to him a plan to prepare Americans for the inevitable outbreak. After that task force meeting, Robert Redfield sent word to Nancy Messonnier, the CDC leader who directly oversaw the failed testing effort, that the messaging was going to shift from containment to mitigation. Redfield, however, neglected to tell her that the shift wouldn't occur immediately.

During the CDC's press briefing the next morning, Messonnier said that the country had crossed the Rubicon. "We expect we will see community spread in this country," she told reporters. "It's not so much a question of if this will happen anymore, but rather more a question of exactly when this will happen and how many people in this country will have severe illness." The stock markets responded accordingly — that is, they tanked.

Trump called Azar the moment his plane landed, around six a.m. on February 26. He was furious. About forty-five minutes before Azar was scheduled to lead that evening's press briefing for the task force, Trump called him into the Oval Office and told him that Vice President Mike Pence was taking charge. Initially, Azar was resigned to this, thinking that Pence's involvement might give the task force more oomph with other Cabinet members.

Out in the briefing room, the announcement, at least at first, went off without a hitch. But as the briefing was coming to an end, there was some confusion among the press about what Azar's current role was. Kadlec watched his boss desperately grab the podium and speak into the microphone. "If I could just clarify, I think you're not getting the point here," Azar said. "I'm still the chairman."

Except . . . not really. Pence's chief of staff now set the agenda of the task force meetings. His principal goal was to avoid the economic turmoil created by a lack of what he called "communications discipline." Azar and his division leaders were instructed not to use the word *pandemic* in the press or say that the situation could "change rapidly." Instead, Azar was asked to pose for photos while leading a prayer group following one task force meeting.

Other players in the White House remained divided about how seriously to take the pandemic. One of them reached out to Deborah Birx, who was in Johannesburg, South Africa, and asked her to join the response as Pence's task force coordinator. Soon to be known for her stylish scarves, Birx had worked in Tony Fauci's lab in the 1980s and later led the military's HIV program under Redfield. She agreed to set aside her work leading the country's global AIDS program in order to occupy this influential position among a group of feuding power brokers in the administration.

The result was that there were now multiple centers of authority shaping the pandemic response. Over the next month, the bruised Alex Azar, leader of the nation's Department of Health and Human Services, became little more than a figurehead perched inside the Hubert H. Humphrey Building, gazing out impotently at the growing outbreak. "It was like that naval tradition where you put the guy on the boat, give them a day of food, and say, 'Good luck,'" Bob Kadlec recalled.

It was a chaotic dynamic that BARDA director Rick Bright stood to gain from. His vision had always been for BARDA to be liberated from the ASPR and elevated to equal status within the HHS hierarchy. During BARDA's annual meeting in October 2019, Kadlec had watched Bright deliver an oddly theatrical presentation. Bright stepped onto the stage and looked out at the audience before enunciating his first word — *"Freedom"* — and letting it hang in the air for an uncomfortably long moment. "The foundation of our nation," he said. "What does freedom mean to you? What is freedom worth?" Kadlec couldn't help but think that for Bright, *freedom* meant freedom from Kadlec's control.

Such independence, it could be argued, was the last thing an operational agency nested in the nation's health department needed during a pandemic. In a lengthy e-mail that Michael Callahan wrote to the Wolverines on February 27, he warned of the feeding frenzy taking place at BARDA. "Mustelids," he began. "Current performers have openly confided this is a great time to grab money from BARDA guppies." The drugs under consideration for funding by BARDA and, to some extent, the NIH were not well suited for a pandemic, he wrote. "In [disaster] operations we select therapies that do not tie up critical personnel (avoid remdosivir [*sic*] if possible)."

Callahan had a particularly jaded view of the pharmaceutical industry and doubted it would deliver a miracle cure. But a miracle cure was exactly what the White House was looking for, and Rick Bright, as it happened, was hoping for the same thing. Bright genuinely felt the urgency of the pandemic, but he also relished the chance for his organization to take a starring role. He soon found an ally in the White House: Peter Navarro. Trump's gruff, upregulated economic adviser was an unlikely bedfellow for the genteel Bright, but the two men started crafting plans to get pharmaceutical companies funded in what Navarro called "Trump time."

At a countermeasures meeting in Kadlec's conference room one morning in March, Callahan learned that Bright wanted to fund a biotech company called Regeneron, located in Tarrytown, New York, to test sarilumab, an intravenous rheumatoid arthritis drug, on coronavirus patients.

Callahan was startled. "What are you funding Regeneron for?" he asked. Sarilumab is a monoclonal antibody that blocks interleukin-6, or IL-6, a signaling molecule the body uses to ramp up the immune response. Regeneron was arguing that in severely ill COVID-19 patients, sarilumab might be able to prevent a cytokine storm, an overreaction of the immune system that causes far more damage than the virus itself.

Callahan had given a briefing for ASPR's staffers on clinical data that he had seen from China, but Bright hadn't shown up. The Chinese had found that only a minority of patients with severe COVID-19 had elevated IL-6 levels. It was one lead to follow, not something to bet the ranch on. But because Regeneron already had an existing federal contract under a special prototyping program, it was easier to extend that contract than deal with the due diligence and arm's-length negotiations required with new contracts. BARDA was going to devote more than two hundred million dollars to support the trial.

After that morning's meeting, Callahan was so upset he went into Kadlec's office and related his concerns, but Kadlec waved him off. These were normal debates, and as much as Kadlec valued Callahan's insights from the field, he didn't think Bright was making a bad call, all things considered. He still hoped to salvage the relationship with the BARDA director and keep his fractured team from falling apart amid whatever came next.

9

THE PRESIDENT'S
POWWOW

On the afternoon of Monday, March 2, President Trump sat at the center of the oval table inside the Cabinet Room with Vice President Mike Pence and Deborah Birx on his left and Alex Azar and Tony Fauci on his right. Executives from ten drug and vaccine manufacturers, including Moderna, Pfizer, Regeneron, and Gilead, were arrayed around the table. The Rose Garden was visible through the window behind the president, the greenery coming to life after a mild winter.

Despite the apparent normalcy of the situation, everyone at the table knew that the nation's first homegrown COVID-19 outbreak was raging at a Seattle-area nursing home called Life Care Center of Kirkland. Four residents in their seventies and eighties had died the previous week and dozens more were experiencing respiratory symptoms. The staff had closed the facility to everyone, including family members, and the police and firefighters who had helped care for the patients who'd died were now under quarantine. Outside the facility, a photographer captured an image of a distraught woman named Bonnie Holstad standing under some trees with a handwritten sign. MY HUSBAND IS A PATIENT . . . KEN HOLSTAD RM 50, it said. NO ONE AT LIFE CARE IS ANSWERING THE PHONES . . . HE NOW HAS A COUGH . . . CAN HE BE TESTED FOR COVID-19? New York, meanwhile, had announced its first confirmed case of COVID-19 on March 1, but even Fauci was still publicly expressing the hope that the outbreak would burn itself out, like SARS had.

Inside the Cabinet Room, Trump leaned into the table, one hand clasped over the other, as he read from a printout. "Today," he said, "we are meeting with the pharmaceutical and biotechnology companies—the biggest in the world, most prestigious, the ones that get down to the bottom line very quickly—to discuss how the federal government can accelerate the development of vaccines and therapeutic treatments for the coronavirus."

During a public health emergency, the FDA was allowed to relax its rules and allow certain experimental products to be used by the general population when no proven alternatives existed. Under the flexible standards of what is known as an emergency use authorization, or EUA, the agency just needed to be convinced that a product "may be effective" and that the potential benefits outweighed the safety risks. The agency had issued EUAs for antiviral drugs during the swine flu pandemic, but it had never authorized a vaccine in that manner in its history. Because vaccines are meant for the healthy, everyone knew the risk-benefit calculus was going to be stacked against an EUA for a vaccine.

One by one, the executives around the table told the president about their capabilities. Stéphane Bancel of Moderna said that his company had finished making its first doses of mRNA vaccine for clinical trials. "We're now waiting for the vaccine to get a green light from the FDA so that the team can start dosing as soon as possible," he said. "We're able to move very, very fast."

"So you're talking over the next few months, you think you could have a vaccine?" Trump asked.

"Correct. Correct. With phase two," Bancel said, referring to the second stage of human clinical trials.

Fauci cut in. "Yeah, you won't have a vaccine. You'll have a vaccine to go into testing."

"And how long would that take?" Trump asked.

"The phase two would take a few months before going into phase three."

"All right, so you're talking within a year—"

"A year to a year and a half," Fauci corrected. The summer or fall of 2021 was the best that could be hoped for, in his view.

"I like the sound of a couple months better, I must be honest with you," Trump said. He continued to go around the table chitchatting with executives, including Pfizer's chief scientific officer, an affable Swedish physician named Mikael Dolsten. Dolsten had secured a last-minute invite to the confab and flew in from Manhattan on the corporate helicopter. Cofounded by German chemist Charles Pfizer, the 170-year-old pharmaceutical juggernaut got its start manufacturing santonin, a drug derived from a rather toxic plant extract that was used to purge parasitic worms from the intestines. The company was now the maker of such blockbuster drugs as Viagra and Lipitor as well as one very successful vaccine, Prevnar 13, which protected against pneumococcal bacteria, a pathogen that can cause pneumonia and other serious illnesses.

Pfizer had announced it was working on a coronavirus drug, but so far it had stayed out of the vaccine race. Like others in Big Pharma, Dolsten's scientists had been lukewarm about derailing their other research to work on a disease still largely confined to China. They wanted to stay focused on, for instance, an mRNA flu vaccine they were developing with BioNTech in Germany. With the coronavirus outbreak showing no signs of waning, however, Dolsten and Pfizer CEO Albert Bourla could see that a course correction might be required. The previous evening, Dolsten had called his vaccine leader, Kathrin Jansen, and asked her to inquire about BioNTech's coronavirus work. The partner company said it was evaluating twenty candidate sequences and beginning to test some of these in mice.

Inside the Cabinet Room now, seated a couple of chairs away from Moderna's Stéphane Bancel, Dolsten kept his cards close to his chest. He spoke of the therapeutics Pfizer was working on and of the company's impressive manufacturing footprint: thirty manufacturing sites in the United States alone, an empire that Bancel could only dream of. Before Dolsten's time was up, he slipped in a carefully worded remark about mRNA technologies that only the most keyed-in observer would have noticed: "They give us an opportunity to move fast," he said. "That's why some of the companies that have been working on other diseases can quickly change priorities and meet the huge public health threat."

Next, Trump turned to Stanley Erck, the CEO of Novavax. How Novavax managed to land itself in the room with so many heavyweights is anyone's guess, but the company had recently added to its roster a new lobbyist who had come straight from the White House. "Thank you for saving the most exciting company for the last," Erck said to laughter. "So we're Novavax. We're down the street in Maryland." He told Trump about the company's technology and the women it had vaccinated in its RSV clinical trial, omitting the fact that the trial had been a failure. He also told Trump about Novavax's new flu vaccine, which was in phase 3 trials. "But we actually—our company is focused on emerging infectious diseases," he said. The company's MERS vaccine had "one hundred percent infection protection" in animals, he said, as did its Ebola vaccine. Erck said that he was expecting results from animal studies within a week, results that would provide the first measure of how potent the neutralizing antibodies produced by Novavax's vaccine were.

"So what do you think in terms of timing?" Trump asked.

Erck said the timeline was similar to what the other companies were talking about, but he added that it was critical that Novavax have a good relationship with the FDA so that "instead of waiting thirty days" for clinical trial approval, it might wait only ten. "And, frankly, we need money," he added. "We're a biotech company, and not one of the larger pharma companies. And so we need money to get to scale."

Trump, who prided himself on being a dealmaker, always seemed most comfortable when surrounded by business types. He told the group at one point that it seemed like they had it all figured out. They knew where they were going, they knew what the answer was going to be, and they just had to get there. A reporter brought up the issue of funding again. "Do you see a need for federal dollars to go to some of these drug companies? I think two of the CEOs around the table mentioned the idea of federal money—" he asked.

"Well, I don't know." Trump's arms were folded across his chest. "I know the companies very well. Some of them are so rich, I think they can actually loan money to the federal government. They don't need money. They need time."

Before the executives departed, Trump invited them into the Oval Office and passed out White House pens featuring his looping signature. Mikael Dolsten of Pfizer hesitated when the president reached out for a handshake, but he decided it would be impolite not to reciprocate. Afterward, he found some hand sanitizer.

• • •

Shortly before three o'clock the following afternoon in Bethesda, Maryland, Francis Collins, director of the National Institutes of Health, joined Tony Fauci in welcoming Trump and Azar to the Vaccine Research Center. In the VRC's seminar room on the first floor, Barney Graham, Kizzmekia Corbett, and John Mascola stood off to the side smiling politely as Collins explained their roles to Trump.

Collins, a slender, gray-haired man with a push-broom mustache, was best known as the gene hunter who'd led the Human Genome Project in the late 1990s. He had a folksy manner about him and often wore a black T-shirt that read VACCINES CAUSE ADULTS when he played guitar and sang in his band, the Affordable Rock 'n' Roll Act. But Collins was a crafty player. He had bipartisan support in Congress and was the rare Obama appointee who had managed to survive the transition to the Trump administration. Collins was a religious man, an Evangelical Christian, but not a conservative, and he appreciated the way in which Secretary Azar had explored the moral and practical issues relating to fetal tissue in federal research and tried to stop the White House from issuing its ban.

With the president appearing to listen intently, Collins made a pitch for his scientists. "People call us the 'National Institutes of Hope,' and we're happy to embrace that particular description," Collins said.

As the discussion turned to vaccines, Fauci took over the proceedings. He walked Trump through the process again and explained how, after obtaining the first SARS virus in 2003, it had taken twenty months to begin phase 1 trials. For Zika, the NIH had gotten that timeline down to just over three months. With the new coronavirus, Fauci said, "I think it's going to be two."

"That's fantastic," Trump said. "That's great."

"So that kind of thing is what this place is all about. It's kind of like a SWAT team going out and responding to emerging microbes." Fauci underscored to Trump the threat that the country was facing while also relating the story of what had happened during the first outbreak of SARS: it disappeared before a vaccine could be tested.

What Fauci didn't mention was that the Vaccine Research Center, in its twenty-plus-year history, had catalyzed many advances but had never delivered an approved vaccine. As accomplished as the NIH's leadership and its scientists were, an academic approach to science and a focus on big issues and big problems was not particularly fruitful when it came to delivering specific products, which was one of the reasons why BARDA had been created. The VRC's leader, John Mascola, had been trained in the military and was determined to score a win for the NIH. Though Mascola was an impressive scientist in his own right, his modesty, mission focus, and interpersonal skills made him the grease that kept the wheels of science turning.

Fauci explained that Moderna's coronavirus vaccine, which had begun as nothing more than Barney Graham's demonstration project, was becoming a critical piece of the government's pandemic response. When he finished his spiel, Trump appeared to be genuinely impressed. "Tony, your reputation is second to none," he said. "I just want to thank everybody at NIH and all of the great scientists and doctors and everything. I know you're working around the clock."

After Trump took a tour of Graham's lab, Fauci and Azar climbed into the back row of "the Beast," the president's armored Cadillac SUV limousine. They were driving to a helipad in order to fly back to the White House. Trump was seated in the row in front of them. "How's Francis Collins doing?" he asked, always suspicious of the Obama holdover's loyalty.

"Francis is great, Mr. President," Azar replied. "You know that fetal-tissue ban you decided on? Francis has been cooperative and professional. He doesn't agree with it, but he's been implementing it faithfully."

"Does that ban affect any of this work getting vaccines?" Trump asked.

"Well, yes, Mr. President," Fauci said. "It does interfere with some of our vaccine and therapeutic work." Moderna was testing its vaccine on cell lines descended from a fetus aborted in the 1970s, cell lines that were not affected by the ban. The use of newer cells, however, was verboten, and researchers could not graft embryonic cells onto mice to better study COVID-19 and the immune system.

"Well, get rid of it, then," Trump said.

Azar, who was not inclined to refight this battle with his nemesis Joe Grogan, told Trump that he should talk to his White House policy staff about it.

10

THIS AIN'T OVER

On the evening of March 3, Grace Fusco was preparing for the weekly family dinner, breading and frying ten pounds of chicken cutlets. "It's not the ingredients, it's the hands," she liked to say. The horse races were blaring from the television, as they always were, and a tablecloth had been thrown over several shoved-together tables spanning the length of the dining room.

Dinner at the Fusco place, a two-story, ten-bedroom home in Freehold, New Jersey, would count as a party anywhere else. The matriarch of a large Italian-American clan, Grace Fusco had eleven children and twenty-seven grandchildren. Her eldest daughter, Rita, came in for the meal, as did her sons Carmine and Vincent. There were about two dozen people in total. Horses were naturally the topic of conversation. Grace's late husband, Vicenzo, had been a trainer and harness racer, driving his prized standardbreds from a two-wheeled cart at the Freehold Raceway, the oldest track in the United States. Carmine and Vincent had followed in his footsteps, becoming trainers, breeders, and racers themselves.

At the end of the night, the Fuscos scattered to their own homes, some of them unknowingly carrying with them the new coronavirus. Carmine experienced a chill that night after dinner, and the next day several members of the family felt under the weather. The locus of the coronavirus crisis had shifted from China to Italy, where the Fuscos still had relatives and where emergency rooms were filling up, but the old country was so far away, the virus so foreign. There were still no confirmed cases in New Jersey and just one in New York. Grace spoke to a doctor, who prescribed her an antibiotic and said things should clear up in a few days.

By then, New Jersey had its first case and, soon after, its second. On March 5, the number of cases in neighboring New York rose from eleven to twenty-two. That day, New York City mayor Bill de Blasio boarded the subway at Fulton Street in downtown Manhattan to demonstrate that residents had nothing to worry about. "I'm here on the subway to say to people nothing to fear, go about your lives and we will tell you if you have to change your habits but that's not now," he said.

The virus was now pressing in on both coasts. A cruise ship called the *Grand Princess* with potentially infected passengers on board was being held off the coast of San Francisco. Kadlec once again sent Michael Callahan to help evacuate the sick. The CDC treated the ship like another breach in containment, even though everyone knew the virus was all around. The virus seemed to be killing 10 to 20 percent of the people it infected, at least among the elderly. Congress, meanwhile, had upped the administration's meager request for a funding supplement, passing a bill to give HHS nearly eight billion dollars to fight the coronavirus. Trump signed it a day later, at 9:05 a.m. on Friday, March 6.

The Fusco family, meanwhile, heard that a gruff but beloved horse-man named John Brennan had died. Brennan, who was sixty-nine, worked as a judge at the Yonkers Raceway and served as an advocate for drivers in disputes with tracks. His death was surprising—Brennan suffered from nothing other than high blood pressure and diabetes. He had developed a fever and a cough at the beginning of March. His condition steadily declined, and he was admitted to the Hackensack University Medical Center, where he was diagnosed with the coronavirus and succumbed to a heart attack on March 10.

On the morning Brennan died, Elizabeth Fusco, the youngest daughter, got a call from her mother, Grace, asking if she would help take care of her horses. When Liz arrived at the house, she found her mother and several of her siblings seriously ill. "I'm not going to the hospital," Grace said. Liz happened to have a pulse oximeter with her for her daughter's medical condition and when she measured her mother's oxygen saturation, she found that it was in the eighties. She took Grace, Rita, and younger brother Joe to an urgent care center. Grace and Rita were sent on to the hospital, but Liz still didn't realize how serious the situation was.

The next day, Wednesday, March 11, marked the day that COVID-19 became real for the world. In a speech that afternoon, the WHO's director general, Tedros Ghebreyesus, officially declared it a pandemic. "*Pandemic* is not a word to use lightly or carelessly," he said, acknowledging those who felt the announcement had been warranted two or three weeks earlier. There were more than 118,000 cases of COVID-19 in 114 countries, and 4,291 people had died. "I remind all countries that we are calling on you to activate and scale up your emergency response mechanisms," he said. "We're in this together, to do the right things with calm and protect the citizens of the world. It's doable."

Even more powerful than Ghebreyesus's speech was the most high-profile COVID-19 diagnosis yet. Actor Tom Hanks and his wife, actress and singer Rita Wilson, posted the news on Instagram with a caption under a photograph of a rubber glove tossed in a biohazard bag. "We felt a bit tired, like we had colds, and some body aches," Hanks wrote from Australia, where he was filming an Elvis Presley biopic. "Rita had some chills that came and went. Slight fevers too . . . Not much more to it than a one-day-at-a-time approach, no?"

Every American was about to feel it. In the United States, a major COVID-19 hot spot had been bubbling up in the New York suburb of New Rochelle, not far from John Brennan's workplace. So many cases had turned up in Westchester County that week—over a hundred by Tuesday—that the state established a one-mile containment zone, shutting down large gatherings days after Hanks's announcement and Brennan's death. Brennan, who never missed a race if he could help it, had had contact with at least one of Grace Fusco's sons not long before that family dinner in New Jersey.

Early on the morning of Thursday, March 12, as Adam Boehler was having an early breakfast surrounded by nautical paintings at the Navy Mess, next door to the White House Situation Room, he received a text message from one of his closest friends: Jared Kushner, Trump's slender, waxy-skinned son-in-law. Involving himself in every issue from the border wall to peace in the Middle East, Kushner was known as the "Secretary of Everything" and when a particular topic captured his attention, you were on his team or you were flattened under his tank tracks. Kushner was from New Jersey and the outbreak was becoming a very big deal in his eyes. *Hey, I need you,* he wrote to Boehler. *Can you come in?*

In contrast to the robot-doll mannerisms of Kushner, Boehler, a wealthy health-care entrepreneur, exuded a chill finance-bro energy. He and Kushner had briefly shared a college dorm room twenty years earlier when they spent a summer working at investment banks in Manhattan. The two bonded over their shared Jewish heritage, and Boehler never forgot the night Kushner had called him after coming back from his first date with his future wife. "I just went on a date with Ivanka *Trump!*" he said. The two were so tight that Boehler had been at Kushner's summer home in 2006 when Kushner's father, Charles, was released from prison after serving a sentence for tax evasion, among other charges. In July 2019, Trump nominated Boehler to lead an organization called the U.S. International Development Finance Corporation.

After Boehler saw the text, he headed across Pennsylvania Avenue to the grand Eisenhower Executive Office Building, where Kushner was meeting with Matthew Pottinger, Trump's deputy national security adviser, Chris Liddell, the White House deputy chief of staff, and some others. All of them were shocked by the way the coronavirus numbers were exploding in the heart of New York City. Governor Andrew Cuomo, the political lion from Queens, had now shut down theaters on Broadway, and school districts would begin closing one by one. Flights from Europe were banned, the National Basketball Association had suspended the season, and the annual St. Patrick's Day parade was postponed indefinitely. "World Turned Upside Down" read the cover of that day's *New York Post*. Kushner asked Boehler how he thought the two of them could best contribute. Boehler said that their strategy should be to find areas where no one else was working on the problem or where it was clear they needed help. Testing was one of the first areas they would dive into.

Kushner reviewed the pros and cons of declaring a national emergency. This was like saying there was a hurricane in every state, and it was something that had never been done in the nation's history. The

White House Coronavirus Task Force members were at a stalemate on the issue, with some advisers fearful of creating more panic and others arguing that they needed to take full advantage of the logistical resources at FEMA, the Federal Emergency Management Agency. With more than twenty thousand employees and regional offices around the country, FEMA had a larger footprint on the ground than the tiny ASPR, and it had greater emergency powers when it came to hiring staff, expediting purchases of medical supplies, and tapping into disaster funds. FEMA could give HHS the support it needed. Kushner advised Trump to move forward with it. At a three p.m. press conference in the White House Rose Garden that Friday, Trump struck a somber tone. Though he didn't say it, he seemed to recognize that the nation was confronting its gravest crisis since September 11, 2001. He vowed to "unleash the full power of the federal government."

Members of the Fusco family were now fighting for their lives. Fifty-six-year-old Rita, who belonged to the local choir, died hours after her admission to the hospital that Thursday. Five days later, Carmine, her fifty-five-year-old brother, died as well. Grace's hospital was locked down to visitors, and doctors wanted to place her on a ventilator. "No," she whispered. Before she would agree to that, she wanted some things brought to her: her rosary beads, her wedding band, and a pillow made from her late husband's pajama pants.

Her daughter Liz hurried over with the items, and she was allowed into her mother's room. The heart monitor beeping in the background, Liz set the pillow at her mother's side, put the beads in her hand. "That was the last time I saw and spoke to my mother," Liz said.

Grace's second son, Vincent, died the next day. Joe, one of the eleven Fusco siblings, spent a month on a ventilator and lost fifty-five pounds before recovering. His sister Maria suffered from semiconscious hallucinations for weeks, terrorized by the thought that her own daughter was dead. At least nineteen members of the Fusco family caught the virus, and five died. Those who survived struggled to make sense of their collective trauma. "This ain't over," Joe warned a visiting reporter. "This ain't over in the least bit."

•   •   •

On the morning of Saturday, March 14, Bob Kadlec descended the carpeted stairs to the basement of his two-story brick home in Alexandria, Virginia. He passed by the picture of himself as a young air force officer and a wall decorated with degrees, plaques, and certificates he had acquired during his long tenure as a public servant. In a storeroom packed with ninety days' worth of military-style meals ready to eat (MREs) and jars of tomato sauce, he dug through old cardboard file boxes filled with yellowing photocopies of obscure internal reports from the military's biodefense research programs.

As newspapers were publishing photographs from China, Thailand, and Korea showing locals wearing face masks, the public message coming from Robert Redfield, Tony Fauci, and other experts was *Move along, nothing to see here.* "There's no reason to be walking around with a mask," Fauci told *60 Minutes* in early March. The thinking was that only N95 masks could filter out virus particles, and those masks were needed by health-care workers. Ordinary surgical masks and homemade cloth masks might worsen the spread of disease. You have to be *trained* to put on and take off a face mask, otherwise you might end up contaminating yourself. The virus was not hanging in the air, the experts said. It was transmitted through close contact — a person coughing next to you on the subway or a deliveryman blowing his nose. Better to wash your hands for twenty seconds, timing yourself by humming "Happy Birthday."

Kadlec believed that the public health advice on this was behind the curve. It all came down to the R_0 — pronounced "R-naught" — the measure of the contagiousness of a virus. The R_0 depends not only on the physical properties of the virus and the environment, such as temperature and humidity, but also on the choices people made amid the pandemic. The best estimate from scientists was that each person who caught the coronavirus was, on average, spreading it to two to four other people. Within a few days, each of those newly infected people would spread it to another two to four people. And each of those people would go on to spread it to another two to four people. And so on. It might not sound like much, but if you began one month with a hundred cases and the R_0 was 2.2, you would have more than 5,000 active cases by the end of that month. At the end of the following month, you would have over

200,000 cases. Kadlec knew that even if a mask was 50 percent effective at blocking transmission, it could potentially cut the R_0 in half. Instead of having 200,000 cases, you might have 200. Over several months of an uncontrolled epidemic, masks could potentially save hundreds of thousands of lives. They would keep hospitals from collapsing, and they would keep the country from having to shut down. How could people not wear them?

Kadlec asked Redfield's team at the CDC to conduct a review and issue guidance on cloth masks for the public. Unfortunately, the scientific literature was mixed on how well cloth masks prevented the spread of respiratory viruses. Masks were hardly a sexy topic for researchers. The last time they had been widely deployed was during the 1918 flu pandemic, back in the dark ages of medicine. But Kadlec knew that's where the nation was headed. They had nothing to stop this thing. After working his way through six file boxes, Kadlec finally found what he was looking for:

TECHNICAL MANUSCRIPT 3
PHYSICAL PROTECTION FROM BIOLOGICAL AEROSOLS
April 1962
U.S. Army Chemical Corps Biological Laboratories
Fort Detrick

This six-page report had been prepared by Bill Patrick, an army bioweaponeer and a mentor of Kadlec's, who had done a fascinating series of experiments. Soldiers ran through puffs of *Bacillus globigii*, a harmless anthrax surrogate, using eight different items to cover their faces: a cotton shirt, a woman's handkerchief, a man's handkerchief, a cotton dress, a rayon slip, a muslin bedsheet, a Turkish bath towel, and toilet paper. Afterward, Patrick swabbed their noses and mouths to see if any bacteria made it in. The Turkish towel folded in half or three layers of toilet paper offered more than 85 percent protection — on par with a surgical mask.

If the country had a tool like this, it could significantly slow community spread right now, Kadlec thought. Unlike the supply chain for N95 masks, the United States had the manufacturing capacity to make cloth masks. He hunted for the phone number that North Carolina senator

Richard Burr, his old boss, had sent him a week earlier for Jerry Cook, an executive at Hanes in Winston-Salem, North Carolina. Kadlec found it, called Cook, and told him that the country needed a favor. The coronavirus might still look like an isolated problem in New York or Seattle, but it was going to be everywhere soon enough. He explained to Cook the deal with the R_0 and the value of masks. *Shelter the susceptible.* "That's all we've got right now," Kadlec said. "It's the only way to slow this thing down until we have a vaccine." He asked Cook if Hanes would make masks for the country.

"How many?" Cook asked.

"We're thinking six hundred and fifty million," Kadlec said.

Over at the NIH, Barney Graham had set his sights on the phase 1 clinical trial of Moderna's vaccine starting by March 10 — sixty days from the release of the WH-Human sequence. It had now been more than two weeks since Kizzmekia Corbett had injected their mice with the second dose of vaccine — plenty of time for them to develop a potent antibody response. Indeed, the antibody-rich blood serum from those mice had prevented their samples from lighting up during the pseudovirus-neutralization assay. The team, in other words, had its first neuts, a confidence-boosting piece of data that Graham had hoped to have in hand before the human trial began. Because meticulous science meant assuming nothing, his collaborator Mark Denison at Vanderbilt was planning to cross-check the pseudovirus assay during the human trial by testing a batch of the same samples against both the pseudovirus and the real coronavirus.

March 10 came and went. Then March 11. Graham asked his NIH colleagues what the holdup was with the human trial. "I wasn't complaining, but I was pressing," Graham said. The FDA was on the verge of approving the Investigational New Drug application, but there were more steps, more paperwork to complete, more meetings to have over Zoom. The NIH was contracting the trial to Kaiser Permanente in Seattle, part of the NIH clinical trial network. The lead clinical investigator still needed to finalize her study protocol and receive approval from her research ethics board.

On March 13, with the president's national emergency declaration, the NIH instituted a partial shutdown, sending home anyone who wasn't involved in coronavirus research or maintaining essential resources. The conference room on Graham's floor was stocked with all the supplies they needed for the next six months and his lab was running three eight-hour shifts around the clock with only two people working at a time. Other laboratories at the Vaccine Research Center that had dialed back their research in other areas helped process the samples coming in from Kizzmekia Corbett's mouse work. Graham's lab was also preparing to ship its immunized mice to Ralph Baric in North Carolina for the challenge trials with the live coronavirus. Over the weekend, Graham settled into his home office, which looked out onto the front porch. On bookshelves filled with academic tomes, he had placed wooden cutouts of the words *faith, hope,* and *love* that his wife, Cynthia, had given him, words that reminded him of the abiding qualities of life. Some might have found them sappy, but not Graham. He was too earnest, too much of a believer. "Without faith, you can't have hope, and without hope, it's difficult to love," Graham said.

The following Monday, March 16, Deborah Birx stood with the president and vice president in announcing a nationwide social-distancing effort, christened "Fifteen Days to Slow the Spread," recommending that people avoid social gatherings of more than ten people and stay at home as much as possible. "We really want people to be separated at this time, to be able to address this virus comprehensively that we cannot see, for which we don't have a vaccine or a therapeutic," she said. Workplaces shut down. City streets emptied out. The State of California went so far as to close its beaches and parks. Even the White House was reduced to a skeleton staff. Anyone meeting with the president had to have his or her temperature taken first.

That day, Graham got word from Seattle that the first four human volunteers, the sentinels, had received their dose of vaccine: twenty-five micrograms of mRNA. One-thousandth of the weight of a grain of rice. Sixty-six days had passed since the coronavirus genome had gone online. If all went well, a total of fifteen people would be vaccinated at the lowest dose, and then the researchers would test doses four and ten times higher. These forty-five volunteers would report any symptoms they experienced after receiving the shots, and their blood would be analyzed to measure antibody levels and the responses of the T cells. Whether or not Moderna's vaccine advanced to the second phase of clinical trials depended, in large part, on these results.

Not long after those first shots, a critical member of Graham's team, Cynthia Ziwawo, developed a cough, and she tested positive for COVID-19 on March 20. She would have to quarantine for over a month, as she remained positive with repeated testing. The protein queen Olu Abiona, who worked right next to her on the lab bench, was terrified of going home and infecting her own parents, both of whom were nurses with medical conditions. She wanted to camp out in the lab, but Graham checked her into a room at the Marriott, where she stayed for two nights until her own test came back negative and she finished her quarantine at home. A third student on the team also went into home isolation, and the lab's preclinical work basically ground to a halt for the next two weeks.

As that mini-COVID crisis was unfolding inside the VRC, Moderna Therapeutics became the company that the nation was pinning its hopes on. The start of its clinical trial on March 16 — the first in the nation for a coronavirus vaccine — sent an upward jolt through the trembling stock market, with the price of the company's shares rising more than 24 percent. Just as this vaccine was going to be a flex for the VRC, it was also a demonstration of Moderna's own capabilities. If Moderna could make it to the finish line with the world's first licensed mRNA vaccine, investors knew, it would help pave the way for the company's entire vaccine pipeline. After ten years of struggle, profitability was finally within reach for Stéphane Bancel's Moderna. Not only had he given up a steady and lucrative job helming a French diagnostics maker for an unproven startup, he had believed in Moderna's mission so much that he sank his own funds into it.

Overnight in Boston, the sidewalks outside the Bunker Hill brownstone Bancel shared with his wife, Brenda, a fine arts photographer, received a dusting of snow. When morning broke the next day, March 17, the serene winter scene was turning to slop. Pfizer, not letting Moderna bask in the spotlight for too long, had issued a press release:

Pfizer and BioNTech to Co-Develop Potential COVID-19 Vaccine

The collaboration aims to accelerate development of BioNTech's potential first-in-class COVID-19 mRNA vaccine program, BNT162, which is expected to enter clinical testing by the end of April 2020. . . . "We are proud that our ongoing, successful relationship with BioNTech gives our companies the resiliency to mobilize our collective resources with extraordinary speed in the face of this worldwide challenge," said Mikael Dolsten, Chief Scientific Officer and President, Worldwide Research, Development & Medical, Pfizer.

As Dolsten tells the story, the timing was mere coincidence. Immediately after returning to New York from the White House roundtable, he had called his CEO, Albert Bourla, and briefed him on his thoughts. Bourla, a veterinarian with bushy eyebrows, had taken the helm of Pfizer just over a year earlier, but they had known each other for many years. Bourla had been with the multinational since 1993, when he joined the animal-health division in his native country of Greece. He was quick on his feet and told Dolsten to go full speed ahead, connecting him with Pfizer's businesspeople. A typical collaboration agreement in biotech would have taken six months to iron out, but Pfizer had a nonbinding letter of intent signed and sent to its partner company in Germany, BioNTech, within ten days. The press release was ready to go immediately.

Bancel had always recognized that Moderna could never stop a pandemic on its own, so the more vaccines racing toward FDA approval, the better for the world. But this development wasn't good at all. His first-mover advantage was going to be threatened by Pfizer's scale and experience. Pfizer excelled at crushing companies large and small, and, through BioNTech's patent portfolio, it shared a license to the same foundational mRNA discovery on which Moderna's vaccines were based. These two rivals were now racing each other on a global stage. Only one could be first.

PART 2

LOCKDOWN

March 2020–May 2020

11

A COMPANY IS BORN

T he path from a scientific hunch to a lifesaving product is rarely a linear one. A decade before the coronavirus pandemic began, a new professor at the Harvard Medical School named Derrick Rossi stepped out of his laboratory at the Immune Disease Institute and carried his laptop to another building on campus. Rossi was a mellow Canadian with a mop of curly hair and a soul patch. It was late in the afternoon on April 27, 2010. Rossi had no idea about the adventure he was about to embark on, but the man he was talking to that day, Timothy Springer, could probably help him out.

Springer was, far and away, the wealthiest scientist that Rossi had ever met. Springer's first company, born out of one of his discoveries in the late 1990s, had led to three FDA-approved drugs. He had been embarrassed by his riches at first, but he eventually embraced his predicament and decided that if he wanted, say, a twenty-three-ton rock in his backyard, he should get it. "I had gone to a presentation on the Chinese tradition of scholars' rock collecting," Springer said. "And I said, 'Aw, that's what I have to have: a rock that is of the same scale as my house.'"

In Springer's office, Rossi presented slides showing a still-unpublished breakthrough that his laboratory had achieved. Using messenger RNA, Rossi and his team had transformed normal cells into stem cells. Stem cells represented the next frontier in medicine. They were like cellular Silly Putty and had the potential to restore a person's vision or heal an injured heart. But the field had been held back because the most potent stem cells came, controversially, from human embryonic tissue, and alternative sources were not ideal. Rossi and his institute had already filed a patent application on their transformational mRNA strategy.

Every time Rossi started to talk about it, however, Springer would stop him with another question. "You expect scientists to be skeptical, but it was a little bit overboard," Rossi said. Despite the interruptions, Rossi continued clicking through his presentation and explained that the potential of mRNA went beyond stem cells. It could turn the body's cells into drug-delivery factories. Once inside the cell, an mRNA sequence would be translated into a protein that could, for instance, counteract a rare genetic disease like cystic fibrosis.

When Rossi finished, he looked over at Springer, whose face had softened. "This is fantastic. I want to invest," he told Rossi. He thought Rossi shouldn't try to do this alone, however. He needed a cofounder, a proven quantity. "Let's call Bob Langer," Springer said.

Langer was a professor at the Massachusetts Institute of Technology and a grand old man of biotech who had founded or been involved in dozens of companies. He had nearly a thousand patents to his name. A month later, on May 25, Langer said he was in. Langer, Springer, and Rossi headed to the offices of Flagship Ventures, where venture capitalist Noubar Afeyan became so excited when he heard about the project that he didn't simply want to bankroll the thing, he wanted to be a cofounder. Screw stem cells, Afeyan said. Let's focus first on making protein therapies for rare diseases. That's the low-hanging fruit and an eighty-billion-dollar market. Rossi's institute was willing to let the four license the mRNA technology described in Rossi's stem-cell patent application.

The nascent company operated in stealth mode at first, waiting for the media blitz that would come when Rossi's results were published in a scientific journal. Rossi mentioned to Flagship's lawyers that his lab's discoveries wouldn't have been possible without the groundwork laid by another biologist, a researcher at the University of Pennsylvania in Philadelphia named Katalin Karikó.

Karikó's unrestrained smile concealed a lifetime of disappointment. She had been working for more than twenty years toward her dream of putting RNA to use as a drug. She was hardly the first to have that dream, but what set her apart was her unwillingness to abandon it. Year after year, she applied for funding from the National Institutes of Health to further her research, and year after year, she was rejected. In January 1995, as Karikó was recuperating from a surgery, she learned that she was no longer eligible for tenure.

She was demoted to a research scientist, and her salary plummeted. She would have to obtain grants to support herself or find other professors to put her on their projects. For Karikó, it was more proof that a woman with a foreign accent could never make it in America. "People always asked me, 'Who is your boss?'" she recalled. "'Who is the head of the lab who you are working for?'" At the Xerox machine on her floor one day, the effusive Karikó struck up a friendship with a new professor, an earnest if reserved immunologist named Drew Weissman. Weissman had come from Tony Fauci's lab at the NIH, and he told Karikó he had been working on a DNA-based HIV vaccine but wasn't having much success.

"Can you make RNA?" he asked.

"Of course I can," Karikó said. Karikó was a confident problem-solver who had grown up behind the Iron Curtain in the dusty farm town of Kisújszállás on the Great Hungarian Plain. Her childhood home hadn't even had running water, much less a television. Her father was a butcher, and his shop was where she first encountered biology; when she was five years old, she learned how blood clots.

Hungary's educational system was stellar, though materials were in short supply during the Cold War. When her experiments required ethyl acetate, a solvent in nail polish remover, Karikó synthesized it by mixing grain alcohol and vinegar and then distilling the combination. In the early 1980s, during her postdoctoral work at the Hungarian Academy of Sciences, she needed small bubbles of fat, known as liposomes, to deliver DNA to cells. To make those, she followed a recipe from the 1940s that included a raw ingredient she knew well from her childhood: cow brains.

With the rise of molecular biology in the 1980s, the concept of gene-based vaccines began to take shape. The biggest hurdle was that DNA on its own didn't generate much of an immune response. One approach sought to package bits of DNA from whatever troublesome virus you wanted to vaccinate people against inside a much less dangerous virus. This was the viral vector vaccine strategy. An alternative strategy — using RNA rather than DNA — didn't seem viable because RNA was considered a relatively unstable molecule. Moreover, as the enzymes needed to make it in the lab didn't become commercially available until 1984, RNA remained challenging to synthesize and challenging to work with — at least if your name wasn't Katalin Karikó.

When Weissman tested Karikó's mRNA in his lab, he found it had the opposite problem that DNA did: It caused too much of a reaction. His cell cultures looked like the aftermath of a bombing raid. He had a hunch about what was happening. To a cell, a strand of foreign mRNA is indistinguishable from an RNA virus. The cell will issue a red alert and stop expressing any and all RNA. A cell that senses it is under attack for long enough will self-destruct. Weissman told Karikó that her mRNA produced such an overwhelming immune reaction, it would not only be challenging as a vaccine, it might never work as a therapy either.

Over the next several years, however, Karikó and Weissman persisted, attempting to circumvent the innate cellular defenses. Organisms of all types generally make mRNA in the same manner, but once that mRNA is released into the cell, the cell will modify it in small ways, adding a new chemical bond here or there. Biochemists have identified over a hundred different types of modified mRNA. For instance, the nucleotide base uracil normally has a sugar stuck to it, forming the mRNA building block known as uridine. One of the most common modifications transforms uridine into pseudouridine. It is the smallest of tweaks, like the way an old-fashioned calculator will add a horizontal bar to the middle of a 0 to turn it into an 8. But in 2004, when Karikó and Weissman tested various types of modified mRNA, they realized they were onto something. The cells no longer blew themselves up. This was huge. Vaccines and drugs using mRNA seemed within reach. They wanted to make a company.

But first, they needed to patent their discovery. Intellectual property is the lifeblood of the biotechnology industry. It is the core of any business plan and the foundation on which to raise money from investors. Researchers at universities in the United States are typically required to hand over the rights to their patents to their employers. Under the terms of their employment agreements, researchers are often granted a share of the proceeds if their technology gets licensed, but their institution gets to make the call on how to commercialize it. Does the university license it out to multiple parties or only one? Should the inventor be encouraged to commercialize the technology independently or is it better to offer it to an established company? Sometimes the goal is to eke out as much profit as possible; other times, it may simply be to do the greatest good for society.

The University of Pennsylvania filed a provisional patent application on Karikó and Weissman's modified mRNA in August 2005. Karikó wanted to develop drugs and thought it would be easy to convince UPenn that the company she and Weissman founded, called RNARx, should be granted a license to the technology. She would finally be able to pay herself a decent salary. UPenn, however, was skeptical. The only way the university would license the patent to RNARx was if the two founders gave the university a majority stake. For every dollar that came in, Karikó and Weissman would be required to turn over fifty cents to the university. Their lawyers told them the company could never survive under such conditions.

On February 5, 2010, the director of UPenn's tech transfer office sent Karikó and Weissman a letter putting an end to their dream. "As we have been unable to reach agreement with RNARx, we have determined to speak with EPICENTRE Technologies Corporation concerning a possible license to the RNA technologies you have developed at Penn," he wrote. "If you are in agreement, please countersign this letter." Karikó was on the verge of tears as she signed it. Epicentre was a small company in Wisconsin that was known for making reagents, not therapeutics. For $300,000, UPenn granted Epicentre an exclusive license. Karikó's initial earnings, after the university deducted its expenses, amounted to $38,250. The university also offered her an adjunct professor position — a job with no path to tenure and a salary of about $50,000 a year.

A few months later, she received a desperate call from a guy in Cambridge, Massachusetts, who said he worked at Flagship Ventures, the firm that was helping bankroll Derrick Rossi's company. "He's telling me, 'I want the patent,'" Karikó recalled. "He was out of his mind," she said. Intellectual property was a Jenga tower — removing a single piece could cause the whole thing to topple. Flagship feared that Rossi's technology — the one Flagship had a license to — likely wouldn't work without Karikó's discovery of modified mRNA. Flagship's start-up, after all, was going to be much bigger than stem cells. It was going to be called Moderna, a portmanteau of *modified* and *RNA*.

But Karikó didn't know much about that yet. All she knew was that there wasn't a lot she could do for this man. Or for herself, for that matter.

Derrick Rossi was about to be put in a difficult position.

The researcher in his lab in Boston who had actually come up with the idea for making stem cells with modified mRNA was a Brit named Luigi Warren. A decade older than the typical postdoc, Warren cut a distinctive figure on the Harvard Medical School campus. He dressed every day in the uniform of a punk-rock producer: black shirt, black blazer, and black jeans. He stood out in other ways as well. For instance, he believed he was being surveilled by the U.S. government and had once warned Rossi that the feds might come after him too. On his blog, Warren championed a convoluted theory that Saddam Hussein had been behind both the 9/11 attack and the 2001 anthrax letters and said that the U.S. government was trying to conceal that "unvoiceable strategic reality." When Rossi told colleagues that he was hiring Warren, their responses were all pretty much the same: *That's the first guy you hired?* The open-minded Rossi brushed off the skepticism.

Warren was frustrated that his paper on stem cells that he hoped would make his academic career was languishing in the editorial pipeline of a peer-reviewed journal called *Cell Stem Cell*. He had been working on the project for over a year and feared that he was going to be scooped by competitors or that someone might steal his idea. Then the editors at *Cell Stem Cell* requested that Warren and Rossi run some additional experiments over the summer. "I'm the one on the front lines," Warren said. "I'm working my ass off here, and I'll lose everything." Rossi was more experienced in the world of scientific publishing. "Guess what, to get it published in a journal, we've got to do this shit," he told Warren.

Warren stormed out of the lab. Rossi had always been able to walk Warren away from the edge when his temper flared up, but this time Warren wasn't coming back. Rossi asked a technician to bring the study across the finish line. After months of work, on August 11, 2010, Rossi learned the good news: the paper was officially accepted. The sweet smell of success dissipated fairly quickly, however. The journal's editor, Deborah Sweet, reached out to Rossi to say that she was putting a pause on the publication because an informant was claiming that Warren's results could not be reproduced. *Who the fuck is replicating our protocol?* Rossi thought. The research was still under wraps, so it had to be an insider.

Next thing Rossi knew, one of the coauthors on the paper, George Daley, a powerful member of the Harvard Medical School faculty, threatened to yank the data that his staff had contributed. Without it, the paper would fall apart. Moderna would be dead in the water. Rossi called Daley and told him he had no right to do that, but Daley refused to budge. Daley's reputation was now on the line. "What a day from hell I have just had," Rossi e-mailed a friend late on the evening of August 26. "Well, can only be uphill from here." Little did he know.

Rossi called the big shots in the department and asked them to broker an agreement with Daley in order to swiftly verify the results. When Luigi Warren heard the news, he marched over to *Cell*'s offices in Cambridge and demanded to speak with editor Deborah Sweet. *This is fucked. I'm going to put a stop to it,* Warren recalled thinking. Sweet told her staff to go home, and Warren was escorted out of the building. Daley soon had a police cruiser keeping watch outside his home in Weston, Massachusetts.

Rossi never learned for certain who that initial informant was or what he or she was trying to accomplish. But over the next few weeks, the stem-cell breakthrough was corroborated to Daley's satisfaction, and on September 30, the article was finally published online. Moderna was covered in the biotech press, and *Time* magazine called the research one of the top ten medical discoveries of the year. George Daley sent over champagne. But the uncomfortable truth for Moderna was that it was founded on nothing — or close to it. Despite the wow factor of the study, the patent that emerged from it was not enough to provide the company with a solid footing of intellectual property.

After Flagship's first inquiry with Katalin Karikó, the company began looking into what it would take to sublicense her technology from the founder of Epicentre, a guy named Gary Dahl. As that process was under way, Flagship floated the idea to Karikó and Drew Weissman of them becoming consultants. It became increasingly clear, however, that Flagship wasn't going to be able to obtain the license for the lowball sum it was offering, a number said to be south of $100,000. The discussions with the UPenn researchers went round and round, and finally Moderna walked away. "The last statement from [Flagship's] lawyer was that they were going to set up the company and spend all their time and money figuring out how to get around our patent," Weissman said.

While the academics saw that decision through a lens of frustration, the businesspeople at Flagship were just being pragmatic. The U.S. Patent and Trademark Office was going to take years to vet Karikó's modified mRNA patent application. Flagship was alert to the possibility that the patent office could narrow its scope, leaving room for Moderna to eke out its own slice of the mRNA patent space. "We didn't know what the patent office was going to do," said Jason Schrum, Moderna's first scientist. The science wasn't even settled as to whether Karikó's mRNA modifications — or any mRNA modifications — were absolutely necessary for the molecules to work as a drug. Alone in a basement in Cambridge, over the course of five months, Schrum tested multiple modified mRNA molecules alone and in combination. Eighty of them, by his estimate.

In early 2011, Flagship hired Stéphane Bancel to be Moderna's first CEO. Bancel, who had trained as an engineer before going to Harvard Business School, was at that time the head of a major French diagnostics company called bioMérieux, but he was eager for something new.

Bancel had grown up in Marseille, France's rowdy, diverse port city on the Mediterranean. His father had been an engineer and his mother a doctor who advised companies. In his youth, he and his younger brother would head south of the city to the mountainous coastline of Calanques National Park, where they would howl and leap off low cliffs into the turquoise waters below. It was a risk, but a calculated one: They would always check to make sure the water was deep enough. When Bancel heard about Moderna's strategy, he could see both its promise and its peril. He knew that RNA was only "stable a few seconds in blood," but he agreed to meet Derrick Rossi at his office to take a look at some of his data.

Bancel came away from the meeting intrigued. For weeks, he weighed the opportunity he was being offered, running it by a number of experts he trusted. During dinner one night at a Beacon Hill restaurant called Bin 26, he and his wife, Brenda, discussed the pros and cons. Bancel was nearing forty and he couldn't imagine a better way to spend the next decade of his life than running Moderna, but he worried about the possibility of failure. "She told me to stop being French and not to be afraid of taking a risk," he recalled to a journalist. The next day, he told Flagship he was ready to take the leap.

Bancel envisioned Moderna becoming the next Genentech, a biotech empire that had an impact in multiple therapeutic areas. He was a demanding boss who lost his temper frequently and was frustrated by the limits of his tiny team. In those first months, he initiated something called Project 800. Its goal was to synthesize eight hundred novel mRNAs, a tedious, overwhelming task without the robots used by Big Pharma. One employee working on the project said she passed out in her shower. Another was suspected of skipping steps in the mRNA manufacturing process, causing some early animal studies to fail. The first head of chemistry walked out of the building in the middle of the day and never came back.

Bancel and Derrick Rossi eventually had a falling-out, and there was further acrimony in those early days — Bancel never suffered fools — but as the company matured, it developed a healthier business culture and hired people better able to keep up with Bancel's demands. "He's not the kind of person who gives up," said his brother, Christophe, also a biotech entrepreneur. Bancel brought on board Stephen Hoge, a handsome doctor and alum from the prestigious consulting firm McKinsey who became the company president. He also hired Israeli doctor Tal Zaks as the company's chief medical officer. Bancel excelled at making sure the company's coffers were fully loaded.

Moderna became a patent-filing machine, building up a fortress of intellectual property, but it couldn't outrun the need for the Karikó and Weissman technology. Their patent was officially approved in October

2012, and it was broad enough that it was going to be hard to build a business on modified mRNA without infringing on it. The following year, Moderna signed a $420 million deal with AstraZeneca to collaborate on an mRNA-based drug to repair damaged heart tissue, along with possible cancer drugs. A breakthrough seemed imminent. With each passing year, the price of the Karikó patent crept upward. The license holder, Gary Dahl, was now unwilling to sublicense the mRNA technology for less than $30 million.

Moderna was unlikely to agree to pay that sum until it had clear path to bring mRNA to the market, and it still didn't have that. In 2016, the company was forced to abandon a clinical trial on what was going to be the company's first therapy, a treatment for a rare disease called Crigler-Najjar syndrome. Patients with the disease are missing a single liver enzyme, and in its most severe form, the disorder can lead to brain damage in infants. As a single-gene disorder that could be treated with only a small amount of enzyme, it seemed like the perfect target. Moderna, however, could never find the correct dose during animal studies. Too much of the drug induced liver toxicity; too little of the drug wasn't potent enough. During Bancel's presentation about Moderna at the J. P. Morgan Healthcare Conference in January 2017, he didn't even mention it. He had turned his attention entirely to mRNA vaccines, which before now had been an afterthought at the company. Bancel understood that a vaccine is given once a year, at best, which meant a smaller payday. But that could also be seen as a plus because the toxicity problems they were wrangling with in their drug pipeline would not be as much of a challenge if a patient received only two or three doses of mRNA over weeks.

That's not to say the shift to vaccines would be easy. In Moderna's first two clinical trials of flu vaccines, a significant number of people suffered from pain and swelling at the injection site, and the levels of antibodies produced were middling, possibly due to the type of fat Moderna had used to stabilize its mRNA and shuttle it into cells. The company honed its approach, and Bancel did his best in the meantime to keep his shareholders happy, boasting of the company's latest intellectual-property acquisition or development. Researchers tinkered away at the formulation of the fats that packaged the mRNA in the vaccines.

In the summer of 2017, Moderna returned to Gary Dahl, whose bargaining position had only grown stronger. Not everyone in the scientific community was persuaded that the Katalin Karikó method of modifying mRNA was necessary, but the alternatives had yet to show results. Moderna agreed to pay seventy-five million dollars for a license to the UPenn patent on top of future royalties.

Karikó earned two million dollars from the deal. By that point, she had been working for three years as the vice president of BioNTech, based in Mainz, Germany. In July 2013, she had traveled from Philadelphia to her native Hungary, where she took a road trip to Switzerland to watch her daughter, a U.S. Olympic rower, compete in the FISA World Rowing Championship. At the time she was entertaining a job offer from Moderna, but it wasn't an executive position. Karikó stopped in Mainz after the championship to give a talk. There, Uğur Şahin, BioNTech's CEO, expressed the kind of warmth she had rarely felt in industry or academia. "We need you," he said. That day, Karikó had a job offer, and BioNTech, like Moderna, gradually changed course from therapeutics to vaccines. It also licensed Karikó's technology under the same terms as Moderna, netting her another two million. At sixty-five, Karikó was grateful for the chance at a comfortable retirement, but the thing she wanted most of all was to see mRNA succeed.

Stéphane Bancel, too, wanted mRNA to succeed. He just wanted to be the first one to do it.

12

RUNNING ON EMPTY

It came so suddenly that no one noticed until everyone noticed all at once. The thing New Yorkers were talking about at first was the toilet paper—or the lack of it. Just like that, it had vanished from the bodega shelves. You couldn't find it at the big supermarkets either. Not only those pillow-soft rolls of Charmin but the cheap, sandpapery single-ply that tears in your hand. Whole aisles of it, gone. *Really? That's the first thing people are hoarding?* the nation wondered. Next went the pasta and rice. Stores soon put limits on the amount of milk and eggs you could buy. One of the few things there wasn't a run on was zucchini.

New York City was soon in a state of panic that hadn't been seen there since 9/11. In the week following the declaration of a national emergency on March 13, 2020, the number of new COVID-19 cases reported in the city spiked from two hundred a day to two thousand. Residents were stocking up on hand sanitizer and wiping down their Amazon packages with disinfectant. They were more suspicious than usual of other human beings: The neighbors in their high-rise buildings. The pedestrians on their sidewalks. Anyone could be carrying the virus. The predominant sound on the streets was ambulance sirens. When emergency doctor Frank Gabrin arrived for his shift at St. John's Episcopal Hospital in Queens on March 19, five ambulances were dropping off patients. That was not normal. "Don't have any PPE that has not been used," Gabrin texted a friend. "No N95 masks."

Over the next few days, Gabrin washed his masks so they could be used for several shifts, and he concocted his own hand sanitizer from vodka and aloe vera. When the only gloves remaining at the hospital were too small, a friend in Florida mailed him some that fit. Facebook and Amazon removed ads for N95 masks, which had been selling at ten times the normal price. Hospital purchasing departments were trying to sequester the last supplies on the open market and leaning on their state health departments for help, which in turn contacted Strategic National Stockpile.

With the HHS secretary Alex Azar sidelined from the task force since late February, various players in the White House had stepped in to manage the operators on the ground. For the doers, working directly for power had its benefits. Those who reported to Vice President Mike Pence or Jared Kushner could get anything they needed at a moment's notice. But as the seat of the response shifted from the under-resourced ASPR to the much larger FEMA, the disaster experts, not the health experts, were making the calls about buying supplies and deploying people and resources. And they were often doing it to satisfy the impulses of their political masters.

Thrust into the middle of this mosh pit was Dr. Bob, who plopped his laptop outside of FEMA administrator Pete Gaynor's eighth-floor office to play second fiddle. He quickly pissed off Gaynor by ignoring White House orders and running his own operations on the side. While FEMA was working with governors to rapidly set up makeshift field hospitals outfitted with porta-potties, Kadlec warned state health officials about fecal spread of the coronavirus. Why not make use of all those empty hotels in city centers? "Tell them to use the goddamn Marriott," Kadlec told his chief medical officer, John Redd, as he deployed to one such hospital inside the Javits Center on Manhattan's West Side. Gaynor called Alex Azar one evening to complain about Kadlec, but Azar told him to buzz off.

By the time New York City's shelter-in-place order went into effect, at eight p.m. on March 22, there were nearly eleven thousand confirmed cases in the city, double the number from two days earlier. All the while, people who had been infected but who had not developed symptoms, perhaps another ten thousand cases, were spreading the virus, embers wafting out from the inferno.

Governor Andrew Cuomo held daily press conferences during which he berated the president for lack of federal assistance and demanded that he invoke the Cold War–era Defense Production Act to force factories to make ventilators under threat of monetary fines and jail time for executives who refused. "I need ventilators, ventilators, ventilators," he told reporters. New York requested fifteen thousand ventilators from the stockpile, which in fact possessed a total of only thirteen thousand at the time.

Kadlec remained steady in the face of all this. Something wasn't adding up. He called Howard Zucker, New York's health commissioner. "I said, 'What's this shit with fifteen thousand ventilators,'" Kadlec recalled.

Zucker said, "I've only got a thousand on hand, and there's modeling that's been shown to the governor that we need twenty times that amount."

"Howard," Kadlec said. "How many do you really have?"

Zucker started going through his list. "I've got seven hundred over here, and a thousand over there," he said.

"You're not even using the ones that you have on hand? How many do you really need?"

"Well, I need a thousand," he said. In the end, Kadlec told Zucker he would send him four hundred from the stockpile and to come back when he needed more. A couple of days later, Cuomo blasted the shipment on television. "What am I going to do with four hundred ventilators when I need thirty thousand?" he said. "You pick the twenty-six thousand people who are going to die because you only sent four hundred ventilators." (His math was off.)

Jared Kushner soon spoke to Cuomo directly and overruled Kadlec, sending an additional four thousand ventilators to New York. On March 27, Cuomo asked for forty thousand. Ventilators were the new toilet paper. "They were hoarding," Kadlec said. "I wasn't even consulted."

The national stockpile needed at least *some* new ventilators. The ones in stock were poorly maintained, and there definitely weren't enough of them if the outbreak spread across the country. Which was why Michael Callahan was offering up his medical expertise on purchases for the stockpile. The disease cowboy proposed funding the scale-up of smaller companies to supply cheap, easy-to-deploy ventilators that could be used just as easily in the back of an ambulance as in a hospital. The team was almost done finalizing a plan when Kushner and his friend Adam Boehler burst into the room at FEMA where the doctors were working and decided to take things in another direction. "We're like, 'Who are these people?'" Callahan said. "The whole system got derailed."

Kushner and Boehler were joined by a team of volunteers they knew from the venture-capital and consulting world who thought they could do the job better and faster than the bureaucrats. They were nicknamed the "Slim Suit Crowd." Boehler made a list of the country's six biggest ventilator manufacturers and decided they were going to lock up their entire inventory for the next six months. These were the Rolls-Royces of the ventilator world—big, heavy machines from Philips and GE meant for fancy, well-equipped hospitals. On March 26, Kushner and Boehler had FEMA issue "notice to proceed" letters obligating the federal government to buy 240,000 breathing machines. Their numbers were based on some real mathematical models coming from the academic world, but they didn't seem credible. They were two to five times higher than what the Brits and the Germans were ordering for their citizens and far more than the most dire estimates of what would be needed in a flu pandemic. "I'm cutting through red tape," Kushner would say. Boehler was looped in on one deal that had the government buying $600 million worth of ventilators from Philips when the company already owed the government nearly identical machines at a fourth of the price.

No one had any clue how they were going to maintain them all, and all the money to pay for this stuff was coming from the HHS piggy bank, which meant that Kadlec and the agency had to sign off on it. "They wanted to do things that were wackadoodle," said one HHS lawyer. And by and large, the White House did whatever it wanted to do.

By the end of March, the crisis in New York was reaching its peak. Frank Gabrin, the emergency doctor, came down with a cough and developed aches in his joints. He told one friend it was probably because he had been using the same mask four days in a row. On the morning of March 31, Gabrin was struggling with each breath. When the paramedics arrived at his apartment, the sixty-year-old doctor was dead.

The challenge at this stage of the outbreak was not just to keep the sick from dying but to figure out how to move the dead. In Brooklyn,

New York, Menachem Bloom, the director of Shomrei Hadas, an Orthodox Jewish funeral home, reached out to volunteers in the community to shuttle bodies to cemeteries. Corpses were lined up on the floor of the chapel, shrouded in white cloth, cinched with red straps. "WE NEED HELP," Bloom begged in a WhatsApp message. "If you have a Minivan or SUV please come and help us. You will get paid for the services." Over a thousand had perished from COVID-19 around the country, nearly three hundred of them in New York City alone. ASPR deployed its DMORTs — disaster mortuary operational response teams — to process bodies at a temporary morgue that the city had set up in a hangar on the Brooklyn waterfront.

Florida's governor signed an executive order requiring anyone who arrived from New York to quarantine for fourteen days. Same deal for Texas, which said violators could face a thousand-dollar fine, a hundred and eighty days in jail, or both. Hawaii asked visitors to postpone their trips for a month, hoping the situation would blow over by then.

For all that, the ventilator apocalypse in New York was a no-show. Doctors learned that COVID patients had better outcomes if they were kept off invasive ventilators, placed in a prone position, and given supplemental oxygen via high-flow nasal cannulas. Kadlec was later able to modify some, but not all, of the contracts that had been pushed through by the White House so that the ventilators could be adapted for this. He was also able to cancel contracts for about ninety thousand surplus ventilators that were far beyond anything the country would ever need. Still, the United States ended up with so many surplus ventilators that it donated many to poorer countries.

Kadlec referred the big ideas raining down from the Slim Suit Crowd and the White House as gifts from the "Crap Fairy." Some of them may have sounded good on paper but weren't practical on the ground amid a pandemic. A drive-through testing program Kushner's team launched at CVS and Walgreens, for instance, sucked up nearly a third of the scarce PPE supplies in the national stockpile. The field hospital at the Javits Center mostly sat empty until its closure on May 1, while a navy hospital ship, the USNS *Comfort*, treated just 182 patients. Kadlec knew from his experience after Hurricane Maria in Puerto Rico that the *Comfort* would never be an operational platform. It was just for show.

Kushner came away from this period with a deep distrust of Kadlec. He told FEMA administrator Pete Gaynor to ignore anything Kadlec had to say, and whenever Kadlec's name came up, Kushner's stolid demeanor would break. One time, he impulsively threw his pen against the wall in a show of his disgust. Another time, Trump mentioned Kadlec during a conversation in the Oval Office. "Kadlec?" Kushner howled. "That's the guy that got us into this mess."

13

MASKS FOR EVERYONE

On the afternoon of Sunday, March 29, Alex Azar was in his wood-paneled office on the sixth floor of the Humphrey Building. The now-powerless secretary had recently shaved off the beard he had worn all winter, perhaps thinking that might give him greater standing with the president, who was not a fan of facial hair. A staffer came in and told Azar that he had a call scheduled with Alex Gorsky, the CEO of Johnson and Johnson. The company had been awarded nearly half a billion dollars by Rick Bright, the head of BARDA, who still, grudgingly, worked for Bob Kadlec. This was the largest federal award yet to a COVID-19 vaccine manufacturer, and Azar was being asked to offer his personal congratulations.

As Azar reviewed a draft of the press release, he saw that the funds were going to support a phase 1 clinical trial and the manufacturing scale-up necessary to produce up to three hundred million doses annually of the company's viral vector vaccine — the kind where the coronavirus spike gene is stitched into the genome of a harmless virus. What Azar didn't see, however, was what American taxpayers were going to receive in return. There was nothing about pricing or purchase agreements if the effort was successful. Azar had already been stung once on this topic. During the annual budget hearing in late February, he had been asked what guarantees the nation could get that coronavirus drugs or vaccines would be affordable. He had blurted out a response about the importance of having the private sector involved. "Price controls won't get us there," he said.

For this he had been pilloried in the media. Azar's commitment to market-based solutions looked out of step during a pandemic that was razing multiple sectors of the U.S. economy. Funding the development of a vaccine that was becoming a national priority required that BARDA adopt a different mindset than it had with far-off threats like Ebola. After his call that evening with Johnson and Johnson's Gorsky, Azar asked Kadlec to come to his office. Azar could get an edge in his voice when he was annoyed, and Kadlec could hear it now. "Bob," Azar said, "how come I didn't know about the J and J deal? We're in the middle of a pandemic."

Beat down after a hard week at FEMA, Kadlec felt defensive at first, but he saw where Azar was going. Both men were struggling to maintain command over their contributions to the pandemic effort. BARDA was one of the last parts of Kadlec's organization that he still had under his control, but even that was slipping away. That week, a divide had been growing between Kadlec and Bright over chloroquine and hydroxychloroquine, two malaria drugs that Trump had declared possible COVID-19 game-changers before any of the data were in.

The cheap, widely available, and relatively safe drugs reduce inflammation in the lungs, so they are often among the first things desperate doctors try when they're dealing with an outbreak of a new respiratory illness — but the drugs rarely work. Bright was initially gung ho on them and got involved in bulking up the national supply of the drugs. Once President Trump got wind of their potential, there was no stopping the chloroquine train. The pushback was immediate — scientists around the country decried the way the unproven drugs were being used with abandon. Bright started to feel uncomfortable with his role in all of this and grew more resentful of Kadlec, who continued to side with the White House on the issue and who was failing to insulate him from the overwhelming political pressure. Kadlec was a physician — or, as Michael Callahan had put it, a human doctor — and the last thing he wanted to do was interfere with a doctor's right to try drugs that were already on the market.

On March 28, the day before Kadlec was called into Azar's office,

Bright had agreed to sign an EUA request to the FDA that would make the drugs available to hospitalized COVID-19 patients, but it wasn't lost on Kadlec and Azar that the BARDA director had been hatching an escape plan. The first element of it came down to the fact that he had convinced his allies in Congress to give BARDA $3.5 billion as part of the Coronavirus Aid, Relief, and Economic Security Act passed just days earlier. Bright would later crow that his slice of the pie represented the first time that BARDA, rather than its parent agency ASPR, got a direct appropriation. Freedom!

Well, not quite. Azar still had the authority to direct how those funds were used, and he told Kadlec that they needed to be strategic about it. With more than twenty division leaders reporting to Azar, he was all about formalizing management processes. On his bookshelf, he had a copy of *The Four Disciplines of Execution,* which he practically knew by heart and which argued that leaders failed to achieve their goals not because of the shortcomings of the people working beneath them but because the systems in place didn't allow workers to see beyond their day-to-day survival. He told Kadlec that when he ran U.S. operations for the drugmaker Eli Lilly, an investment of fifty million dollars on a clinical trial had to be reviewed by both the global board of directors and a science committee. Azar said the same thing should hold for BARDA. He proposed creating what he called the "Secretary's Science Advisory Council" to review Bright's portfolio. He and Kadlec would bring in the big dogs from across HHS, people like NIH director Francis Collins, FDA commissioner Stephen Hahn, and CDC director Robert Redfield. Bright groused to one of his supporters that Azar and Kadlec were trying to tie his hands.

On Sunday, April 5, the Secretary's Science Advisory Council held its first meeting over videoconference. Bright clicked through slide after slide and droned on about this drug or that vaccine. "He's slide-whipping us," Azar texted Kadlec. Bright was moving so fast and with so little detail on BARDA's portfolio that none of the experts had a chance to chime in, but the slide-whipped participants already had concerns. BARDA had funded Regeneron's expensive trial on sarilumab—the interleukin-6 inhibitor that Michael Callahan had been so incensed about —as well as another IL-6 inhibitor from Genentech. None of this had been fully coordinated with the NIH, which was running its own adaptive trials with multiple drugs, including an IL-6 inhibitor. Collins and Tony Fauci, also at the videoconference, both found BARDA's approach scattershot, lacking balance and systematic prioritization. It did not seem as if Bright had a coherent plan, much less one primed to deliver vaccines in a public health crisis.

The next day, Kadlec was due at the White House Situation Room to explain to the vice president's task force what on earth he was doing talking to the U.S. Post Office about cloth masks.

For the past three weeks, Kadlec had been nailing down the details of his nationwide mask effort, dubbed America Strong. "Keep Healthy and Live On" was the motto. Trump had already touted Hanes's involvement in the effort during a briefing in March, and Kadlec had the first prototypes sent to his house in late March. His wife, Ann, and daughter Samantha were standing in the kitchen when he came in with a few samples and asked them to test them out. Samantha put the white stretchy underwear fabric over her face and tried to adjust it to fit under her eyeglasses. "I'm going to send them to every family in America," her father said, beaming.

Robert Redfield was telling the news media, finally, that asymptomatic transmission was occurring. The CDC was preparing to issue new guidance recommending cloth face coverings to the general population. When Kadlec spoke to Deborah Birx, the White House Coronavirus Task Force coordinator, about the masks, her only request was that they come in a neutral color. Not only would Kadlec's cloth masks take the pressure off the medical supply chain, they would be a powerful tool to limit community spread of the virus.

With its five-hundred-million-dollar contract from HHS, Hanes had shifted all of its production lines in the United States from making briefs to making face masks, and it was churning them out seven days a week. The company reopened operations in the Dominican Republic, El Salvador, and Honduras and sent Hanes-made textiles to other manufacturing partners working on the effort. The first masks off the line went straight to Hanes employees, and the others were packaged in bags with a printed label and a USPS tracking number. Boxes were soon stacking up inside USPS distribution centers and the warehouses of the Strategic National Stockpile. The USPS crafted a draft press release touting the "historic delivery" in partnership with the White House Coronavirus Task Force. "We stand ready to deliver, as we have for 240 years, to help our country combat the pandemic," the postmaster general was quoted as saying.

Jared Kushner and Adam Boehler thought that Kadlec's low-tech masks were a dumb idea and a waste of money. They looked like jock-straps, said one task force member, which, truth be told, they kind of did. Some feared they would panic the country. Russ Vought, the direc-tor of the White House budget office, didn't want to give any extra busi-ness to the U.S. Postal Service, which conservatives had been trying to starve to death.

Inside the Situation Room on April 6, Kadlec was about to present the details of America Strong to the task force when it was struck from the agenda. Kadlec sat dumbfounded. Afterward, Vought told Kadlec by phone to cancel the contract with Hanes and the other companies, but it was too late. If Kushner could force HHS to buy a bunch of use-less ventilators, Kadlec thought, then he would damn well buy these masks. Kadlec resolved to distribute them to urban health clinics, pris-ons, and other underserved communities. But as for a mask delivered to the doorstep of every American and a coherent message about the pub-lic health benefits of masking? That wasn't happening.

If the missteps in February and early March could be distilled down to denial, confusion, and miscommunication, then the month of April marked the stage where ideology and impatience took over. This was when Trump suggested at a press conference that we might be able to kill the coronavirus by shining a "very powerful light" into the body or, per-haps, by injecting disinfectant for a "cleaning" of the lungs.

Then there was the never-ending controversy over chloroquine and hydroxychloroquine, drugs that Trump wouldn't shut up about. The emergency authorization Rick Bright had requested from the FDA was reversed after the drugs proved useless and, it was feared, potentially harmful. In hindsight, Kadlec said, Trump touting miracle cures "was like Lyndon Johnson picking bombing targets in Vietnam."

It was also in April that Deborah Birx told the White House that the outbreak had peaked in hot spots like New Orleans and Houston and that a gradual decline across the country was possible if social distancing continued into the summer. On the *Today* show, Fauci lowered his estimate of the death toll of the coronavirus over the next five months from 200,000 or so down to 60,000. Again, that hinged on social distancing being respected. The administration paid attention to the first part of their message but not the second. The pandemic had gone on long enough, was the growing view, and it was time to reopen for the sake of the economy. If state governors wanted to slow things down, that was on them.

Going forward, the White House saw the real fight as no longer against the virus but against the media.

Enter Michael Caputo. As Caputo tells it, on March 26 he was eating lunch in his kitchen in East Aurora, a small town near Buffalo, New York, when his cell phone rang. The caller ID was a string of zeros. It was the president, asking if Caputo wanted a job. "I need somebody I trust," Trump said.

A Grateful Dead fan who wore a skull ring on his right middle finger and woven bracelets on his wrists and ankles, Caputo had grown up between his mother's mobile home in Ohio and his father's place in Buffalo. A longtime campaign operative, he had been mentored in politics by the dirty trickster Roger Stone, and he'd driven Trump around when he came to visit the capital. Caputo saw Trump as a disruptor in Republican politics who might do something for the workingman.

Now Trump, without alerting Alex Azar, was going to make Caputo his man inside the Humphrey Building as the HHS assistant secretary for public affairs. Caputo's job would be plugging leaks to the media and making the administration look good at all costs. Caputo, whose public relations firm had been struggling and who'd been planning to apply for a federal small-business loan, packed his bag and drove the seven hours to Washington, DC. He reported for his first day of work on Monday, April 13.

Days later, the NIH's Kizzmekia Corbett came under attack from the right for her tweets decrying the toll the virus was having on the Black community. A relatively high proportion of the Black population couldn't work from home because they had jobs in health care or food service, and many Black families lived in multigenerational homes, which created a greater risk of spreading the virus to older relatives. The Black population overall was also in poorer health, suffering from greater rates of obesity and diabetes than the general population, which made them more at risk for developing severe COVID-19. In Chicago, Black people made up a third of the population, but they represented more than half of all COVID-19 cases and two-thirds of the deaths. Hispanic and American Indian communities were also suffering disproportionately.

"Some have gone as far to call it genocide," Corbett wrote. "I plead the fifth." Fox News host Tucker Carlson ran a segment on his program on April 17 arguing that she should be focused on "questions of science" not "people's skin colors" and "some weird race theory." He concluded by saying that if these were the kind of people working for the NIH, then it's "not going to be a useful scientific body much longer because it's lunacy." One of Caputo's deputies told Fox that ethics officials at HHS would be looking into it.

Meanwhile, after Tony Fauci disagreed with the president one too many times, the White House press shop blocked him from giving television interviews — at least until public outcry about his absence became too great. "Trump's now back in charge," Jared Kushner told journalist Bob Woodward that month. "It's not the doctors."

As for Bob Kadlec, he could see that his career, and the country, had reached an inflection point. With this White House, you were either in or you were out, and Kadlec was most definitely out. Worse, he had been given the job he had waited for his whole life, and he had let the nation down. On the evening of April 6, back at home after the task force meeting, he thought about Azar's disappointment with him over BARDA's go-it-alone dealmaking. He also thought about Fauci's pessimistic (but realistic) timeline for vaccine approval: twelve to fourteen months. An eternity in a pandemic. Dr. Bob did have more to offer the country, it turned out. But as he had learned during his Special Operations days, when you were trying to survive in a hostile environment, it was better to move under cover of darkness.

14

THE MOONSHOT

Bob Kadlec was standing in his office holding a dry-erase marker, the words *Manhattan Project* scribbled across the top of a whiteboard. It was the afternoon of Friday, April 10, and he was surrounded by four of his most trusted associates and a very large stack of pizza boxes. The group had assembled here for a historic brainstorming session. At exactly five p.m., Kadlec said, a man by the name of Peter Marks would be calling. Marks was the director of the FDA's Center for Biologics Evaluation and Research (CBER), the referee who would make or break the fortunes of vaccine companies.

The two men had come to know and respect each other from previous conversations about pandemic preparedness. One was a former spy, the other a geeky doctor, but they had an undeniable chemistry. That morning, they had briefed Alex Azar on an ambitious plan they had hatched during six a.m. calls over the past few days. They were setting their sights on a Manhattan Project for vaccines. Their program would be headquartered inside the Humphrey Building, but, like the original Manhattan Project—the secretive American effort to produce the first atomic bomb—this new plan would bring all the might of American industry and government to the task.

This was exactly the kind of strategic thinking that Azar had been demanding from Kadlec, and to have Kadlec bring him *the guy* from the FDA enthusiastically saying it could all be done in record time—Azar was sold.

A string bean of a man with tortoiseshell glasses, Marks was an oncologist by training. Following a successful but not entirely rewarding career developing drugs at Novartis, he had entered government service in 2010 after seeing an FDA job advertisement in the *New England Journal of Medicine*. Soul-stirring stuff, right? He was a career employee who was supposed to be insulated from politics. He soon found his calling: figuring out how to turn the hazy answers that science provides into the go/no-go decisions that regulators and the industry need. In the past month, Marks had been having conversations with drugmakers, including Pfizer and Johnson and Johnson, who seemed to him to have a "defeatist attitude" when it came to the COVID-19 vaccine timeline. Johnson and Johnson was talking about 2022 as a realistic target. Marks was nothing like your stereotypical regulator who said, *No, no, no.* He was more like, *Maybe, huh, yes, eureka!*

When he'd first reached out to Kadlec a few days earlier, he and his team at the FDA were already thinking about how the agency should evaluate an application for an emergency use authorization for a vaccine. The legal standard was vague. How much evidence were they going to require? Under normal circumstances, when a vaccine candidate entered human clinical trials, regulators in Marks's office evaluated it every step of the way. As Barney Graham and his team at the NIH demonstrated earlier, to begin a first-in-human phase 1 trial, you have to show that a vaccine is safe and has the potential to produce an immune response in animals or cell cultures. If Marks's office was convinced from the results of the phase 1 trial that a vaccine was likely to be safe and possibly effective, it authorized the company to move into phase 2 trials, which might involve several hundred volunteers. These trials determined the most common side effects and homed in on the appropriate dose. If a vaccine still looked good after that, it advanced to the pivotal phase 3 trial.

A phase 3 trial represents the gold standard in the scientific process. It is the ultimate demonstration of whether the vaccine prevents infection, disease, or death. Unlike clinical trials for many drugs, which need to recruit tens or hundreds of sick patients to determine if the treatment helps them recover over the course of days or weeks, a clinical trial for a vaccine requires thousands or tens of thousands of healthy volunteers who are followed over months or years. The volunteers are randomly assigned to receive either the vaccine candidate or a placebo. To avoid any bias from creeping in, neither the company, the researchers, nor the patients know who receives what until the trial reaches a predetermined end point and the data are unblinded. The reason a phase 3 vaccine trial often lasts for a year or more is that it takes that long for members in both the vaccine and placebo groups to be naturally exposed to the virus during their day-to-day activities. The two groups are then compared with a statistical test to determine if the vaccine provides a health benefit. Such a large trial also offers a chance to identify rare side effects that didn't show up in the earlier phases. At the end of this yearslong process, Marks's team reviews all the raw clinical data submitted by the company, along with quality-control tests of its manufacturing process, and it approves a vaccine only if the FDA believes the benefits of it outweigh any potential risks.

The long-standing problem with running a trial during a disease outbreak is that the outbreak often wanes before enough cases of disease or death occur in the placebo group. The summer lull that some models were predicting for COVID-19 would be bad news for a phase 3 clinical trial. Results might not roll in until late fall 2020, and by then, it would be too late to stop the tidal wave of deaths that would come with rising case numbers. What's more, scaling up the manufacturing process doesn't normally start until a vaccine is approved, leading to more months of delay. What to do?

Inside the ASPR conference room that historic Friday, the phone rang, and Marks's squeaky voice came over the speaker. Marks was, at that moment, pushing his cart through the aisles of the Whole Foods in Bethesda, Maryland, on his weekly socially distanced shopping trip. During their discussions that week, Kadlec had put a challenge to Marks: Sketch out exactly what it would take to issue an EUA as quickly as possible using any available regulatory pathway, perhaps even without the benefit of phase 3 data. Now Marks laid out a scenario where they would run a more intensive set of animal studies than was usual for standard vaccine development. He wanted studies with multiple species: mice, hamsters, ferrets, and monkeys. The immunology results would then be cross-checked with the phase 1 safety trials in humans. This would allow greater confidence regarding which animal species were the best models for how humans would respond to the vaccine and suggest whether deploying a vaccine without the completion of a phase 3 trial was going to be an acceptable risk.

But Marks, in an inspired riff, took this idea farther, to a place not even Kadlec had anticipated. Marks realized that those animal studies could be part of a rigorous competition to evaluate the candidates of a sea of companies vying to make it big with their coronavirus vaccine. Since the government was funding the whole thing, they wouldn't just focus on the Mercks and Pfizers of the world; they'd look at underdogs with a promising approach. The government would pay to ramp up manufacturing for those top candidates even while the vaccine was still being tested, so any that were granted the green light could start going into the arms of vulnerable adults immediately. Simultaneously, two or three of the most promising candidates would continue to be evaluated in an efficient, head-to-head phase 3 study known as a master protocol. If everything went according to plan, Marks suggested, a vaccine could be rolled out by December 2020.

"How about October first?" Kadlec asked, tearing off his second or third slice of a pizza. He was famished.

Marks wasn't going to split hairs. As they spoke, the regulator became amped up, envisioning all the agencies working together "in the truest sense of a team," as he put it. He felt that there should be close collaboration not only among the U.S. agencies but also with Richard Hatchett at CEPI, the international vaccine effort. This was a kumbaya moment, "a time," Marks told Kadlec, "for countries to come together."

"I don't know about that, Peter." The military man jumped in with his realpolitik. Governments were going to look out for their own citizens first. That's what happened with vaccines during the 2009 swine flu pandemic, and it was bound to happen again. National security always came first. More to the point, Kadlec wanted the effort to be small, agile, and mission-focused, not mired in bureaucracy or wrangling over datasharing. America had the money and resources to move faster than any international effort.

Chandresh Harjivan, a pharmaceutical expert whom Kadlec had invited to offer his advice, brought up the cost of a program that would scale up vaccine manufacturing before the final tests had been run. "Money is not a concern here," Kadlec said, pointing out that more than two thousand people were dying every day from COVID-19 and that the nation was losing tens of billions of dollars every day it was shut down. Azar, the business-minded conservative, had told him there was a $3 trillion return for the country on a successful vaccine.

Recognizing the national security importance of the project, Azar and Kadlec would enlist the Department of Defense as a key partner in both the science and logistics. They wanted to call the effort Manhattan Project 2.0, or MP 2.0, but both Marks and Harjivan scoffed at naming a public health initiative after a program meant to create destruction. Marks, a *Star Trek* fan in his youth, said that he was calling his component of the effort, the animal studies, Project Warp Speed.

Over that weekend, Kadlec and Marks drafted a memo and a set of slides that Kadlec would share with Azar. Kadlec circulated the materials, including a four-point mission statement, to colleagues at six p.m. on Sunday, April 12.

Project Warp Speed

Maximally expediting a safe effective vaccine

A safe, effective, broadly-administered vaccine is the single most important solution to COVID-19 pandemic

MISSION: **Maximally expedite** the development of a safe and effective vaccine with sufficient scale to inoculate all Americans who need it

DEADLINE: Enable broad access to the public by **October 2020**

PLAN: Modeled after the **Manhattan Project** approach, a **multi-disciplinary, multi-sector team** that brings the numerous in-flight efforts under a **single authority** to drive relentless coordination, barrier elimination, and accountability for mission success

Kadlec knew that there was only one path that the secretary could take to convince the White House to approve MP 2.0, and that particular path involved Jared Kushner. Which meant Kadlec couldn't be a visible part of it. The more Kushner saw of Kadlec, the more likely Kadlec was to be fired. So while Kadlec served as Azar's inside man at the Humphrey Building, Paul Mango, the HHS deputy chief of staff for policy, became the White House interface. He was a polished McKinsey guy, an operator who could smooth over any turbulence. Unlike Kadlec and Azar, Mango fit in at the White House in another way as well: he rarely sported a face mask.

Early on the morning of Wednesday, April 15, Azar and Mango headed to the West Wing to talk to Kushner and Adam Boehler about new strategies for sustaining the Strategic National Stockpile in the long term. As the meeting was coming to a close, Azar presented the concept for MP 2.0 as though he were offering it to the president's son-in-law to spearhead. Kushner's eyes lit up. He opined that Azar's pharmaceutical-industry experience made him the obvious person to oversee such a program. Kushner would do whatever he could to help him, but in addition to expediting vaccines, they should add therapeutics to the mix. It was exactly what Azar had been hoping to hear.

Kushner and Boehler went upstairs and caught President Trump as he arrived at the Oval Office from his executive quarters, where he typically received his infusion of Fox News each morning. They told him that Azar had brought them this idea for a Manhattan Project for vaccines. They knew that the speed of the program, light-years ahead of what the president had been hearing before, would appeal to him. How was that possible? Trump asked, given everything that Tony Fauci had been saying. They explained that manufacturing the vaccines while they were still being tested would be costly, but it could shave six months or more off the timeline. "Get it done," Trump said.

Back at the Humphrey Building, Kadlec focused on the operational details of MP 2.0. He was a student of history, but he realized he didn't know exactly how the original Manhattan Project had worked. He asked his personal historian — well, an ASPR consultant who had been researching bioweapons in the National Archives — to write up a two-page brief for him. It described how the orderly effort had marshaled the nation's scientists to work in lockstep in order to transform "atomic energy from a theoretical issue in 1939 to a practical problem with a solution in sight in 1941."

In reviewing this history, Kadlec came to a striking realization. MP 2.0 didn't need one leader. It needed two. "We need our Robert Oppenheimer and our Leslie Groves," he told Chan Harjivan one day. Oppenheimer was the physicist who had served as scientific lead; Groves was the lieutenant colonel in the Army Corps of Engineers who took charge of construction efforts and the acquisition of raw materials. They needed both to win this war. "Who is going to be acceptable to Trump?" Kadlec mused.

For MP 2.0 to be successful, all of HHS, including the Centers for Disease Control and Prevention, the National Institutes of Health, and the Food and Drug Administration, would have to come together and work as a single operational unit. The leaders of MP 2.0 would drive the mission forward with what Kadlec's historian had called a "simple, clear objective," and those leaders would best be able to achieve that goal if they came from outside the health agencies and their entrenched cultures.

"You should lead it," Harjivan said.

"No friggin' way," Kadlec replied.

On the afternoon of April 16, Kadlec received a phone call that would threaten his plans for MP 2.0. It was NIH director Francis Collins. Collins was working on a novel idea to develop rapid diagnostic tests for COVID-19. It was called RADx and was like a scientific version of the reality show *Shark Tank*, where a panel of experts put candidates under intense scrutiny in a concentrated time frame and guided the winners into commercialization. Collins, however, had been unable to persuade Rick Bright to support the program with any of the $3.5 billion that Congress had given BARDA. Collins told Kadlec he needed around $250 million and someone who was good at working with industry to get the diagnostics scheme going.

Not long after Kadlec hung up with Collins, his phone rang again. Two Republican senators who led the health committee, Roy Blunt and Lamar Alexander, were on the line and so angry at Bright for not helping out that they wanted to take that whole $3.5 billion and transfer it to NIH. "Mr. Secretary, you can choose to support this or not, but we're going to be speaking to the president this afternoon," one of the senators told Kadlec.

There was no mistaking it for anything other than what it was: a threat. Kadlec couldn't afford to create further animosity, not while he needed all the stars to be aligned for MP 2.0 to get off the ground. Perhaps, Kadlec thought, he could keep both Collins and the senators happy by sending Bright over to the NIH along with a billion dollars? Yes, that was the ticket. Collins's RADx rapid testing program could be the third prong of MP 2.0 alongside the therapeutics and vaccine efforts.

The next day, another wrench was thrown into his plans. Bright remained bitter about his role in the shenanigans surrounding one of Trump's "game-changer" drugs, chloroquine, and he was even more resentful of the way Kadlec and Azar had been meddling in his affairs. Kadlec got called into a secret meeting with Azar's team and learned that Bright had leaked to a reporter internal e-mails related to the administration's efforts to fast-track donations of chloroquine coming from questionable factories in India and Pakistan. Michael Caputo, finishing his first week as the president's leak-plugger in the Humphrey Building, thought Bright should be canned for making the administration look bad. "Are you kidding me?" Kadlec said. "No!" If they fired Bright, Kadlec said, he would be forced to send Collins someone like Gary Disbrow, one of the linchpins at BARDA he needed for vaccines. Kadlec said that Bright should just be reassigned. It was decided to do it quietly over the weekend, shutting down Bright's e-mail before he had a chance to copy and leak any more documents from HHS servers.

A couple of hours later, Kadlec called up Bright and asked him if he was familiar with the phrase *catastrophic success*. In light of the national emergency, he said, Bright was about to be given his greatest responsibility yet, leading the NIH diagnostics shark tank. Bright was perplexed. "Can I still be the BARDA director?" he asked.

"This is a full-time endeavor," Kadlec replied.

In Bright's first call to Collins to talk about the diagnostics plans, Collins laid into him. He told Bright that he knew he was a difficult employee and that everyone was angry with him. As Collins later recalled, he was frustrated. "This was not my idea," he said. "This guy has some-how caused the secretary to lose confidence in him and now I'm sup-posed to take him?" Collins told Bright he was doing him a favor by let-ting him come to NIH and work for him on this effort. Bright was taken aback. He wasn't working *for* Dr. Collins. He was working *with* him. He was going to be the one leading the effort!

When Bright told Kadlec about the call, Kadlec didn't conceal his anger with the NIH director. "Jackass," he muttered indelicately. He was growing awfully tired of everyone putting themselves before the mis-sion. *Suck it up* was his philosophy, and his alone, apparently.

That Saturday, Bright became worked up when he discovered that his name and picture had been removed from the BARDA website. He sent Kadlec several messages, but Kadlec didn't respond until the next day. "I've confirmed you've been assigned to NIH," he texted on Monday, April 20. Bright's e-mail account was at last cut off, but not before he se-cured copies of more department e-mails.

It was already shaping up to be a rough week for Alex Azar. He had attended the burial of his eighty-year-old father the previous week, and now the *Wall Street Journal* was apparently working on a bombshell about his lack of leadership in the early days of the pandemic. Two reporters sent over a list of topics they wanted to ask Azar about, and he was apoplectic about their line of questioning. He called — who else? — Michael Caputo into his office. Caputo and Azar were in a shotgun marriage of sorts. Azar hadn't been happy about the way Trump foisted Caputo on HHS, but he was not blind to the benefit of working with a power player. "This is Joe Grogan fucking me!" Azar said. "And I do not want to get fucked."

Azar was convinced that Grogan, the head of the White House Do-mestic Policy Council, was somehow orchestrating the hit piece. Gro-gan took on larger-than-life proportions in Azar's descriptions and Ca-puto wasn't sure if Azar was paranoid or onto something, but he knew he couldn't pick sides in this apparent death match. He needed to con-tain the fallout of this battle and plug the leaks for the president's sake. "If you can't kill a story, you can hobble a story," explained Caputo, the practiced media relations hand. He told Azar that they would tape the *Wall Street Journal* interview. When the story came out, they would take control of the narrative by pointing out everything the reporter failed to quote Azar on. Fake news, they would call it. Trump's base would eat that stuff up.

On Wednesday, April 22, the *Journal* published its devastating piece, arguing that Azar had tried to cover up the CDC's testing foibles and had left HHS in a shambles.

Then, in another rotten development for Azar, it emerged that Bright had been talking to a civil rights lawyer named Debra Katz whom the *Washington Post* had called "the feared attorney of the #MeToo moment." Before Katz tried to take down Supreme Court justice nominee Brett Kavanaugh over sexual-assault allegations, she had fought for Bright's partner through a whistleblowing complaint related to gay discrimination during the George W. Bush era. Now Bright was planning to fling his own Molotov cocktail at the Trump team. He had been pushed around for too long. He was going to take aim at the most vulnerable target, his own supervisor: Dr. Bob.

In a statement Katz released to the media on April 22, Bright claimed that his transfer to NIH was a retaliatory response "to my insistence that the government invest the billions of dollars allocated by Congress to address the COVID-19 pandemic into safe and scientifically vetted solutions, and not in drugs, vaccines and other technologies that lack scientific merit." He said, "I am speaking out because to combat this deadly virus, science — not politics or cronyism — has to lead the way."

Hunkered down in the Humphrey Building, Bob Kadlec was getting down to brass tacks. He and Peter Marks were looking into commandeering the nation's monkey-research capacity as part of Marks's full-on animal-testing agenda. On Wednesday, April 22, they had a call with military researchers, including people from the army's top biodefense laboratory in Fort Detrick, Maryland, to discuss their plans. Afterward, Kadlec looked over at Marks and said gravely, "It's not a good time to be a nonhuman primate in this country." That got a few laughs.

An unprecedented scramble for lab animals was also playing out in the private sector. Moving those clinical trials from the safety stage to the larger efficacy stages required data from monkeys. China, which supplied more than half of all primates imported to the United States, cut off all exports in April as part of a wildlife trade ban and to preserve its own coronavirus research programs. If your COVID-19 research wasn't funded by the NIH, you had little chance of gaining access to rhesus macaques, the preeminent research monkey, in their colonies.

Down in Texas, a former army virologist named Tom Geisbert saw the price of rhesus monkeys double on the open market. "It felt like it was an auction, man," he said. "You'd be talking to a vendor and he'd be like 'Hey, I got twenty rhesus over here, and I've got a guy who'll give me seven thousand, you want to go eight, eight, eight. Going once — ten!" Novavax obtained its baboons from Texas. In Rockland, Maryland, Mark Lewis, president of Bioqual, a pay-for-play lab, turned to cynos — cynomolgus macaques — for his work with Johnson and Johnson and Moderna. He even bought so-called dirty monkeys, which meant they had been born outside a sterile facility. "We were stuck taking whatever we could get."

This was the beginning of the boom times for Gilbert "Sully" Gordon, a monkey trapper living on the Caribbean island of St. Kitts. The island had a colony of more than sixty thousand feral African green monkeys who lived like pirates, stealing mangoes and smashing watermelons on local farms. These miscreants are about the size of large housecats, with dark, triangular faces and long skinny tails. Sully had thirty of his living-room-size traps deployed all over the island. He headed out on his ATV each day, baited his traps with twenty pounds of sweet potato, and bided his time until the monkeys got too comfortable with the easy life. Snap!

One of the benefits of those African greens was that they developed more of the serious clinical signs of COVID-19 than did rhesus macaques, who showed little more than the sniffles. The MP 2.0 team planned to run vaccine trials with African greens at Southern Research in Birmingham, Alabama, and at Lovelace Biomedical Research Institute in Albuquerque. Other labs around the country would be testing mice, hamsters, ferrets, and rhesus macaques when they could obtain them.

Working closely with BARDA, Marks was narrowing down the list of ninety-two vaccine candidates the MP 2.0 team was considering. The Defense Department, for instance, had been funding the biotech firm

Inovio, which was developing a DNA-based vaccine that required a special device, known as a gene gun, in order to be injected. Marks didn't think it was practical, but he wasn't quite ready to write it off. "Beam me up, Scotty," he said, adding it to the portfolio.

On Friday, April 24, nine days after Trump gave the effort his unofficial approval, Marks e-mailed Kadlec to see if it was okay for him to start talking directly to companies about the Warp Speed concept. "We need to pressure-test who could be ready for animal studies by June first and phase one studies by early July."

Kadlec tapped out his response on his iPhone: "Go man go."

The program had fully captured Marks's imagination. That Saturday morning at his home, Marks pieced together a logo for his Project Warp Speed, overlaying the image of a syringe atop the arrowhead-shaped shield from the *Star Trek* franchise. His wife, an artist, suggested he make up buttons, and he spent twenty bucks to order a hundred of them to give out to his future Warp Speed team.

After a week of bad press for Alex Azar, Jared Kushner asked his friend Adam Boehler whether the HHS head should be fired. Boehler weighed the options. He told Kushner he could go either way, honestly, but Azar wasn't the root cause of the problem. The problem was that Azar had been disempowered. "We can't be in purgatory here," he said. "Either you fire Azar and you bring somebody else in, or you reconstitute Azar." Kushner spoke to the president shortly afterward and then told Boehler to go forward with the plan to remake the secretary.

Boehler asked Azar and his deputy chief of staff, Paul Mango, to come to his home that Sunday, April 26, with a strategy for the remainder of Trump's term. He lived on DC's verdant Foxhall Road in a rented fifteen-million-dollar mansion previously occupied by the French ambassador. Boehler's children greeted Azar, and he and Mango headed into a lounge overlooking the pool in the rear of the house. Boehler jokingly called it the "men's lounge," in contrast to a brighter space on the other side of the home. A few Keto Bars had been laid out on a table for Azar, who suffered from celiac disease.

Another Kushner ally, Brad Smith, set up an easel pad, and the four men went through a five-point plan. With the hurricane season coming up, day-to-day management of the pandemic response would have to be transferred from FEMA to the HHS, but they would install a Coast Guard admiral to run the show, limiting Kadlec's responsibilities and leaving Azar to focus on his top priorities. Azar kept returning to the importance of being protected from his enemy Joe Grogan, but Boehler kept directing him back to MP 2.0 and his drug-pricing policies. Focus on that, and the rest, Boehler said reassuringly, would sort itself out.

Azar wasn't so sure. On Tuesday, he read a front-page story in the *New York Times* about the Oxford University group's viral vector vaccine program pulling ahead of the competition. He hadn't heard anything about the Oxford group's work before, and he felt blindsided. *Oh, crap,* he thought, his survival instinct still on high alert.

Over the past few weeks, the Oxford team had been championing its program in the media. Rather than running a small safety trial, as Moderna had, the Oxford team hit the ground running by recruiting more than a thousand participants in a larger test of a single dose of its vaccine. Sarah Gilbert said that she was 80 percent confident that Oxford's vaccine, called ChAdOx1, would be proven effective by September. Gilbert's colleague Adrian Hill went further in one interview, saying that they were going to be a full year ahead of their competitors. "Do you have any concern that you're being overly optimistic, that that just seems, for lack of a better word, too good to be true?"

"We don't think so," Hill replied.

The *New York Times* article picked up on the thread. "The Oxford scientists," the story read, "now say that with an emergency approval from regulators, the first few million doses of their vaccine could be available by September—at least several months ahead of any of the other announced efforts—if it proves to be effective." Azar knew that if he didn't move fast, he would get an angry call from Trump. He told Kadlec he needed to head that off by figuring out how the vaccine could be manufactured in the U.S.

Oxford, it turned out, was in the middle of signing a ten-million-dollar deal with the British drugmaker AstraZeneca. AstraZeneca wasn't a major player in the vaccine business, but it was a global company and willing to move swiftly and also to help supply poorer countries with vaccine. The potential price of the AstraZeneca vaccine at mere dollars per dose and the fact that it might require just a single dose gave it a leg up over the mRNA vaccines, which could end up costing ten times as much. Azar contacted Matt Hancock, the British health minister, to find out how to be a part of the Oxford effort, but he wasn't out of the woods yet. Far from it.

Michael Caputo was walking around the Humphrey Building with his cell phone glued to his ear. At 1:45 p.m. that day, April 29, Azar was slated to brief Joe Grogan's domestic policy council on what exactly HHS had been doing about therapeutics and vaccines. Azar was tense about the meeting with his nemesis, and he and Caputo had been readying a preemptive strike in the form of a positive news story. Caputo had given an exclusive about MP 2.0 to a reporter he trusted at Bloomberg News. A "Joe Friday" type, he called her. As in, "Just the facts, ma'am."

But that morning Caputo was also communicating with Maggie Haberman, the shrewd *New York Times* White House reporter. Haberman and Caputo had known each other for many years, and she was a formidable foe. She and two other reporters were working on another big piece about Azar that suggested he was still on thin ice. The Rick Bright imbroglio was just the latest episode indicating that, at best, Azar was failing to properly manage his people and, at worst, that HHS was going to botch vaccine development the way it had diagnostic testing. Haberman wanted a statement. "You know, Maggie, I'd be real careful," Caputo warned her. He dismissed the gist of the story. It was yesterday's news. Azar was going to be just fine.

Caputo was now even more eager for Bloomberg to publish its piece before Team Haberman rained on their MP 2.0 announcement parade. In the hallway outside the secretary's suite, he caught Peter Marks and Bob Kadlec. Caputo told them he had wanted to name MP 2.0 Project Apollo, but that plan had to be nixed after Joe Biden, Trump's opponent in the upcoming presidential election, spoke about an "Apollo-like moonshot" for vaccines.

"Did you have any other names?" Caputo asked.

"Operation Warp Speed," Kadlec said, giving a nod to Marks. With

the military involved, it could no longer be just a "project." Caputo repeated the name. He liked it.

"They're going to think we're all nerds." Marks sighed.

Over at the White House, Azar led Marks downstairs to the Situation
Room, and Azar took his assigned seat across from Deborah Birx, the
task force coordinator, on the long side of the table. Marks got stuck on
the back bench. Bob Kadlec tuned in from the secure communications
line at the Humphrey Building.

Once the meeting was under way, Marks piped up from the back of
the room for the run-through of his slides. Grogan stopped him and
gave him a seat at the head of the table and then Marks began again,
describing how he and his team had come up with a scheme to prioritize fourteen vaccines using a five-point scale under five different criteria, including the status of development and the likelihood of success
with a particular manufacturing platform. At the top of the list, with
fourteen points, he ranked the vaccine Merck was developing using the
same platform as its Ebola vaccine, which had received FDA approval
six months earlier. This coronavirus vaccine was a chimera; it would use
a weakened strain of the vesicular stomatitis virus outfitted with a functional coronavirus spike that did the dirty work of infecting cells. Unlike
the Oxford vaccine, which generated a onetime burst of spike proteins
inside cells, the viruses in Merck's live-attenuated vaccine made copies
of themselves, producing multiple generations of spike-covered viruses.
Next up was Sanofi's protein-subunit vaccine, which earned thirteen
points. Tied for third place were the mRNA vaccines from Moderna and
Pfizer, with ten points each. Johnson and Johnson and Novavax were in
fourth place, while the Oxford vaccine, with just five points, was tied for
last place with an obscure company in California. About half of those
vaccines had already received some support from the federal government, but Marks had a few long-shot candidates on the list, including an
oral vaccine made by a company called VaxArt.

He then clicked through to a chart showing the parallel development
pathways he envisioned, with comparative animal trials, at-risk manufacturing scale-up, and human clinical trials all happening at approximately the same time. According to his calculations, ten million doses
could be available under an emergency use authorization by October

2020 with a goal of three hundred million doses with full approval by January 2021. He was aiming for 80 percent efficacy for the vaccines. The team would select its scientific leader within a week.

At the end of his presentation, Deborah Birx scowled at Marks. "These are all unproven platforms," she said. There were no approved vaccines for any disease that used the viral vector technology or mRNA. How could he possibly consider Moderna a front-runner? As for the protein-based vaccines developed by Sanofi and Novavax, supposedly the sure bets, they both came from the same fussy caterpillar-cell platform. Birx never forgot the debacle of VaxSyn, the HIV vaccine that she had tested as a therapeutic with Robert Redfield back in the 1990s. Never mind that the platform had come a long way in thirty years. No, Birx said, they should be growing proteins in Chinese hamster ovary cells, a method used for rare-disease drugs. "Why don't you get Genentech to make the vaccine?" she said.

"Genentech doesn't make vaccines," Marks said wanly.

Then Mark Meadows, Trump's new chief of staff, who was sitting directly behind Azar, spoke up. "How much have you made already?" Meadows asked.

"We've just started the program," Marks said. "We're going to start testing—"

"I can't believe this," Meadows said. "Why are we waiting? You need to be manufacturing this at risk. You should have been doing this months ago."

Joe Grogan watched this exchange with great interest.

As Azar left the disastrous meeting, Tony Fauci grabbed his arm. "Trust me, Alex, Moderna is going to work," he said. "We will have a vaccine." Marks stormed out, upset that Azar had failed to defend him, the political rube. Outside, he took one of the bright red bikes from the Capital Bikeshare racks at Seventeenth and G Streets. He pedaled up Pennsylvania Avenue toward his home in Georgetown, blazing through several red lights, as it dawned on him that he had basically been led, unprepared, into a war zone in the White House. He was still panting when Kadlec's consultant Chan Harjivan called to ask how the meeting went. "I just biked home," Marks said, "and I didn't wear a helmet."

"Why not?" Harjivan asked.

"If I got hit by a car, I'd probably be better off." When Marks stepped inside his home, he opened up a link to the Bloomberg News story, which had gone live at 2:08 p.m. "Trump's 'Operation Warp Speed' Aims to Rush Coronavirus Vaccine," read the headline. It was really happening. Maybe the day wasn't so bad after all.

15

RETURN TO POWER

Alex Azar left the Situation Room after the meeting and learned that Maggie Haberman's negative story about him had come out in the *New York Times*. Azar thought it had Joe Grogan's fingerprints all over it. He found Adam Boehler in the lower lobby of the West Wing waiting around for another meeting to start and told him that he would resign as secretary if he had to keep enduring Grogan's attacks. The president should know that Grogan was a leaker. That despicable man was a destroyer, not a creator, he said. What Azar and his people needed more than anything was to be protected from this bullshit. Hours later, Grogan announced that he was resigning from his position as head of the Domestic Policy Council—a plan, Grogan told reporters, that had long been in the works.

With one less enemy to worry about, Azar was riding high on his way to the Pentagon two days later, on Friday, May 1, for a 7:30 a.m. meeting. The headquarters of the Department of Defense is situated on the Potomac River across from the capital in Arlington, Virginia. The security escort Azar's driver was following got lost winding through the maze of roads surrounding the building. Eventually, they found their way to the ramp leading to the defense secretary's private entrance.

Once Azar was inside, a military aide and several assistants guided him through the outermost ring of the Pentagon, the E-ring. They passed the wall of portraits of former defense secretaries on the second floor and came to the large, nondescript Nunn-Lugar Conference Room across the hall from the secretary's suite. A coffee machine was set out on a rolling tray. Several tables were arranged in a triangle. Adam Boehler was already there on one side. On the other side sat David Norquist, the deputy secretary, and Mark Milley, the chairman of the Joint Chiefs of Staff, in his army service uniform decorated with half a dozen badges and multicolored service ribbons. Azar sat along the third side with his chief of staff. After a few minutes, the defense secretary, Mark Esper, came in. Jared Kushner was the last to arrive.

During their first in-person meeting, the assembled power brokers laid out how the DoD would give Operation Warp Speed the logistical support it needed to move supplies and distribute vaccines. They discussed the organizational structure, and everyone agreed that it was critical that the Warp Speed team be self-contained, a single unit placed inside a protective cocoon to minimize the involvement of the feuding factions in the White House. "We want to keep it clean," Azar said. "We want to keep it out of the grips of the task force."

But because Operation Warp Speed was engaging multiple branches of the government, the White House would have to have some nominal form of representation, and the men agreed that Kushner and Boehler would serve on a board of directors with Azar and Esper. Later, they would add Deborah Birx as a safeguard to prevent the vice president's task force from subsuming the effort. Francis Collins, Tony Fauci, and Robert Redfield would become nonvoting advisers to the Warp Speed board.

So who was going to lead this thing? The men sketched out a picture of the person on top, the Leslie Groves character who would need to manage both the daunting number of supplies required for the vaccines and, ultimately, the distribution of doses. Esper told Milley, "This sounds like a Gus Perna job."

Gustave Perna was a four-star army general, one of only forty-five in the nation holding the military's highest rank. Perna led the U.S. Army Materiel Command, headquartered at Redstone Arsenal outside of Huntsville, Alabama, and he was on the verge of retiring. An imposing man with a superhero-size torso that seemed like it was about to burst out of his camouflage fatigues, he had deployed to Iraq three times in his career, commanding resupply operations. He was a Jersey guy who never met a football analogy that he didn't like. Fluke passes, end zones, and handoffs were all part of his lingo. "I'm executing zone offense and zone defense," he said at a Q and A event earlier that year. "I'm trying to cover the whole spectrum knowing that it is significantly bigger than my capability and I'm just trying to outmaneuver the enemy." What Perna lacked in scientific knowledge, he made up for in instincts and his ability to distill a complex problem into exactly what he needed to know to make a life-and-death decision.

Across the country in downtown Seattle, Ian Haydon wasn't the least bit worried about the risks he was taking. It was Tuesday, May 5, and he was receiving his second dose of Moderna's mRNA vaccine for the phase 1 trial. He walked past the shuttered Starbucks in the lobby of the Metropolitan East Building and took an elevator up to the sixteenth floor to Kaiser Permanente's Research Institute. A month earlier, inside a sparsely furnished room in this building, Haydon had signed a twenty-page consent form warning him of the potential dangers he faced. Being a clinical trial volunteer was no joke. But Haydon had heard good things about the Moderna vaccine. He worked as a science writer for the University of Washington, and a professor there had told him that the mRNA vaccine was "super-cool" and seemed likely to work.

When Haydon had his first shot, he'd barely felt a thing. A prick of a needle in his deltoid. He was given the highest dose planned for the study: one-quarter of a milligram, or 250 micrograms. He waited in the clinic for an hour so that staff could ensure he didn't have an allergic reaction and then he ordered Thai food to be delivered to his house as a small celebration. He'd gone home with a thermometer and a diary to record how he felt over the next week, but there wasn't much to report.

Now he was back at Kaiser for the second dose. He felt some pain develop in his arm after an hour but otherwise felt fine. Around ten p.m., he got chills and put on a pair of sweatpants and a sweatshirt. "Is it cold in here?" he asked his girlfriend. She told him it wasn't. He was feeling tired and crawled into bed. At one a.m., Haydon stirred and felt like his head was about to explode. He was boiling hot, soaking in his own sweat. He checked his temperature and saw it was 103.2 degrees. He put the thermometer back on the nightstand and tried to sleep it off, but he was in pain and felt nauseated. His immune system was out of kilter. Around four a.m., Haydon found the sheaf of papers he had been given at Kaiser and called the twenty-four-hour hotline. They told him to head over to the hospital immediately. One of the things running through his mind at that particular moment was *What is this going to mean for the vaccine?*

Haydon recovered over the next day, and Moderna dropped this highest dose from its clinical tests. Haydon's reaction helped show that the dose exceeded the maximum amount of mRNA that the company could reasonably give a healthy person. What the NIH's Barney Graham and others rooting for the Moderna vaccine were desperate to know was whether a lower dose of vaccine would generate a sufficient antibody response. If a lower dose failed, the likelihood of ending the pandemic anytime soon would decrease immeasurably. If it succeeded, well, there were still a lot of other things that could go wrong.

That's why it was so critical for the Warp Speed team to settle on its Oppenheimer, the technical lead, who could work with the companies and oversee the experts at the various health agencies. They needed an industry insider, someone who could drive the effort at an unrelenting pace.

In the second week of April, Jim Greenwood, a former Republican congressman and the head of the Biotechnology Innovation Organization, suggested a guy named Moncef Slaoui to Alex Azar. Slaoui, a soft-spoken Moroccan-born scientist with the square jaw of a fighter, had been head of vaccines at GlaxoSmithKline. Slaoui had shepherded fourteen of the company's vaccines through approval, a record that no one could compete with. So revered was Slaoui at his former company that it named its Maryland vaccine center after him. Since his retirement, he'd been serving on the board of Moderna, where he had gained a deep knowledge of its mRNA technology and developed the ability to work with Stéphane Bancel, with whom he shared a mother tongue. Azar passed the name on to Kadlec, and he started looking into Slaoui.

When Kadlec reached Slaoui to chat about the possible job, Slaoui told him that he was willing to begin immediately, and he wouldn't take a penny from the government. "What's motivating you?" Kadlec asked.

Slaoui had fallen into the business world in the United States, and he'd made a boatload of money, but had he truly served his adopted country? Not in the way his father had served his own people when he was imprisoned for fighting against the French occupation in Morocco. Slaoui had gotten into vaccines to save his countrymen and now he had a chance to do that. "I'm a naturalized citizen," he told Kadlec. "This is my duty." Kadlec told Azar that Slaoui was his first choice—he was the Oppenheimer.

He wasn't the front-runner at that stage, however. The official interviews for Warp Speed's technical lead were taking place on May 6. The Warp Speed board, minus Deborah Birx, all arrived at the Humphrey Building. Adam Boehler, Paul Mango, and Tony Fauci went into one conference room; Jared Kushner, Alex Azar, and David Norquist, the deputy defense secretary, went into another. The teams sized up each of the seven candidates for thirty minutes. On paper, the favored candidate seemed to be Elias Zerhouni, the former NIH director, but when he was asked whether the program could succeed, he replied, "Perhaps not." That was an immediate red flag. You don't send a general who thinks he's going to lose into a battle. A second candidate who seemed promising: Sanofi's former CEO Chris Viehbacher was based in Germany and wanted to lead the effort from afar. As for Slaoui, one of the strikes against him was that he had never worked in government. But, he observed, that was in fact an upside. "I don't owe anybody anything," he said. "All I'm trying to do here is get to the finish line as quickly as possible."

After the interviews, Slaoui was everyone's first choice, but his biggest problem was that he was a walking financial conflict of interest. He got paid for being on the boards of companies that Warp Speed was going to give hundreds of millions of dollars to, including Moderna and its manufacturing partner Lonza, and Slaoui also held ten million dollars' worth of stock in his former employer GlaxoSmithKline. To take the top spot, he would have to obtain an ethics waiver. Even so, he would surely become a political liability.

Kushner argued that this wasn't a reason to go with one of the timid bureaucrats farther down the list. They decided that Slaoui would have to be brought on not as a government employee but as a consultant, a position that would trigger less strict ethics rules. Slaoui would also have to agree to donate any gains in his stock portfolio to the NIH over the course of his term. General Perna would become the official lead, the chief operating officer who had to answer to Congress if anything shady happened.

On May 13, Slaoui's black Ferrari roared into the parking of the Humphrey Building. When Michael Caputo went in to greet him in a conference room on the sixth floor, Slaoui had the look of a man at ease in his position, sporting designer jeans and a black leather jacket. Some papers were spread out on the table in front of him, and he was tapping out a message on his phone.

Caputo told Slaoui that if any reporters called about his position, he was to say nothing. Slaoui was going to give an exclusive interview the next day to Maggie Haberman of the *New York Times*. Once that was done, he would be sequestered to focus on his work. "You're going to get kicked in the balls once, maybe twice," Caputo said matter-of-factly, as people do in Buffalo.

"Kicked in the balls?" the ever polite Slaoui replied with a look of puzzlement.

"Yeah, kicked in the balls. About your stock portfolio," Caputo said.

"Why? I'm going to lose millions over this," Slaoui said.

Around eleven a.m., Alex Azar took Slaoui over to the White House to meet the president. Defense Secretary Mark Esper had come with General Perna from the Pentagon. Jared Kushner brought them all into the Oval Office together. President Trump looked admiringly at Perna in his brown four-star-general dress uniform. Esper briefed the president on Perna's military career and made a point of noting that he was from New Jersey, which he knew Trump would appreciate, as so much of his real estate empire had been built there. The president was awed by the initiative and excited by the idea of the military carrying out the distribution. He thought that meant they would be driving vaccines around in Humvees, though Esper clarified that they'd be bringing in commercial shipping companies for that.

"These are the best people on earth," Kushner told his father-in-law. "We are going to get the vaccines we need."

"Can we really get a vaccine in a year?" Trump asked Slaoui.

Slaoui told him it was doable, but there were no guarantees in the vaccine business. They were going to give it their best effort. Trump said that was fantastic, just fantastic.

The next day, May 14, at ten a.m., Anna Eshoo, a California Democrat, pounded her gavel inside room 2123 of the Rayburn House Office Building. She was leading a hearing called "Protecting Scientific Integrity in the COVID-19 Response." Eshoo had been the one who had helped Bright secure his BARDA appropriation, and she had invited him to testify before the nation. She expressed disappointment that Azar and Kadlec had "refused to testify today" and asked "for a moment of silence in honor of the over eighty thousand Americans who have lost their lives from COVID-19."

With a flag pin on his lapel and his gray hair meticulously combed to the side, Bright leaned forward and closed his eyes. All was quiet in the room but for a few clicks of a camera. He then removed his face mask and read his testimony. "Today, the world is confronting a public health emergency unlike any we've seen in over a century," he said. "Without better planning, 2020 could be the darkest winter in modern history."

Bright's legal team had publicly released its sixty-three-page whistleblower complaint, painting Bright as the lone hero inside the ASPR. He had gone from being a big supporter of the agency stocking up on supplies of chloroquine and hydroxychloroquine that emergency department doctors were clamoring for to crying foul over the way the White House wanted to push the unproven drugs out to pharmacies. His attacks were not focused on the president's misstatements, however, but on Bob Kadlec, who, he said, was a Trump flunky who posed a "danger to public health" and had abused his authority, broken the law, wasted funds, and censored scientific research for his own corrupt ends. That whole chloroquine thing? That was Kadlec's fault. The N95 mask shortages? Kadlec's fault as well.

Bright also got his payback on Michael Callahan for the dressing-down Callahan had given him in February. Bright alleged that Callahan and Kadlec had skirted BARDA's "rigorous" review process in order to run a clinical trial on famotidine, a generic heartburn drug that Callahan's Chinese colleagues had suggested might work against COVID-19. "This contract was just one more example of Dr. Kadlec's actions in bypassing all rules and procedures designed to ensure public safety and to avoid corruption in the award of billions of dollars in government funds," Bright's complaint read. Kadlec was already a black sheep inside the Trump administration and now he was also a villain to the world. Callahan told Kadlec he blamed himself for Bright. "I helped create this monster," he said. "I brought him to government when he was nothing."

Over nearly four hours of testimony, Bright repeated the charges in his whistleblowing complaint, claiming that Kadlec had set the vaccine effort back. He also pointed out that Kadlec had failed to properly ramp up supplies of remdesivir, the drug Bright had hounded Callahan about back in February. As Callahan had predicted, remdesivir wasn't saving lives or even reducing the amount of virus in a patient's respiratory tract. It did seem to shorten hospital stays from fifteen days to eleven, but that modest effect had been demonstrated only after Bright left BARDA.

Eshoo asked, "Instead of acting on your recommendations, was the response of others to try and cut you out of key meetings, marginalize your participation?"

"I was told that my urgings were causing a commotion and I was removed from those meetings," Bright said. Later, he pointed the finger directly at Kadlec for that failure. "I think my removal was because of tensions and actions I took between my supervisor, Dr. Kadlec, and myself." Although there was a certain truthiness to the charges Bright leveled about the administration's anemic response to the pandemic, even some of Bright's own admirers felt that what had unfolded at the ASPR was a personality conflict tucked inside a pressure cooker. It was bound to explode.

At one point during the hearing, Bright was asked how long he thought it would take for the nation to have a vaccine. "A lot of optimism is swirling around a twelve- to eighteen-month time frame if everything goes perfectly. We've never seen everything go perfectly," he said. "I think it's going to take longer than that to do so."

That afternoon, Alex Azar was in his motorcade heading out to an event in Allentown, Pennsylvania, with a supplier for the Strategic National

Stockpile. He was almost to Baltimore when he got a call. Change of plans: Trump wanted him back immediately. The president was fuming about the media coverage of Bright, and he wanted to drown it out with news of his own making. Azar would join him at the White House and they would announce Operation Warp Speed and then head to the stockpile together to show how committed the president was to fighting the pandemic.

Azar's vehicle exited I-95 and boomeranged back to the capital. When he arrived at the White House . . . another change of plans. The press team decided the announcement should wait until the next day. Trump, however, would still head out with his health secretary. The two men left the Oval Office and crossed the White House lawn to board Marine One, the olive-green helicopter that would ferry them to their destination. As they passed the gaggle of reporters, Trump told Azar to "hit back hard" on Bright.

Azar didn't hold back. His face was flaming red as he blasted the whistleblower over the roar of the helicopter engines. "Everything he is complaining about was achieved," he said. "He said we needed a Manhattan Project for vaccines. This president initiates a vaccine Manhattan Project, diagnostic Manhattan Project, therapeutic Manhattan Project. Oh, and by the way, whose job was it to actually lead the development of vaccines? Dr. Bright. So while we're launching Operation Warp Speed, he's not showing up for work to be part of that. So this is like somebody was in a choir and is now trying to say he was a soloist back then. What he was saying was what every single member of this administration and the president was saying."

While the secretary caught his breath, Trump sputtered a few remarks about Bright being "an angry, disgruntled employee" who "didn't do a very good job." Then they boarded Marine One and the president sank into his plush gray seat. A grin hung on his doughy face. Azar felt a sense of relief. Rick Bright's atom bomb had been defused. The center of power had been restored to the Hubert H. Humphrey Building.

PART 3

TESTING

May 2020–October 2020

16

HORSE RACE

It was with no small measure of anticipation that Peter Marks rode his mountain bike to the White House on May 15 for the official launch of Operation Warp Speed. The nerdy regulator locked the bike up in front, passed through a security checkpoint, and grabbed one of the ten folding chairs spread out on the lawn of the Rose Garden. He nodded at Bob Kadlec, who was sitting quietly about as far as he could get from the action.

At 12:40 p.m., President Trump emerged from the White House in a navy suit, flanked by Alex Azar, Francis Collins, Tony Fauci, Deborah Birx, and Mark Esper. "This is going to be a very hot one, and we apologize to everybody out there that's going to suffer through it," Trump said, squinting in the sunlight. It was one of those days when he had an inhumanly orange complexion from the makeup he slathered on in ungodly amounts. "Tomorrow will mark thirty days since we released the White House guidelines for a safe and phased opening of America. That's what we're doing. It's the opening of America. We're going to have an amazing year next year."

Trump described the central pillars of his reopening plan: the ramping-up of diagnostic testing and the acceleration of drug and vaccine development. "Scientists at the NIH began developing the first vaccine candidate on January eleventh — think of that — within hours of the virus's genetic code being posted online," he said. "Then my administration cut through every piece of red tape to achieve the fastest ever, by far, launch of a vaccine trial for this new virus, this very vicious virus.

"Today I want to update you on the next stage of this momentous medical initiative. It's called Operation Warp Speed. That means big and it means fast. A massive scientific, industrial, and logistical endeavor unlike anything our country has seen since the Manhattan Project. You really could say that nobody has seen anything like we're doing, whether it's ventilators or testing. Nobody has seen anything like we're doing now, within our country, since the Second World War. Incredible.

"Its objective is to finish developing and then to manufacture and distribute a proven coronavirus vaccine as fast as possible. Again, we'd love to see if we could do it prior to the end of the year." Operation Warp Speed would also encompass Francis Collins's diagnostics shark tank and a program to accelerate development in therapeutics, including monoclonal antibodies and the antibody-rich convalescent plasma from recovered COVID-19 patients.

Trump gestured toward Moncef Slaoui, whose forehead showed a sheen of sweat. "One of the most respected men in the world in the production and, really, on the formulation of vaccines," Trump said. "When a vaccine is ready, the U.S. government will deploy every plane, truck, and soldier required to help distribute it to the American people as quickly as possible."

Slaoui approached the microphone. "I have very recently seen early data from a clinical trial with a coronavirus vaccine," he said, referring to a confidential preview of Moderna's phase 1 results. "And this data made me feel even more confident that we will be able to deliver a few hundred million doses of vaccine by the end of 2020."

For a nation that had been hearing for months from reputable scientists that a vaccine couldn't possibly be rolled out until the summer of 2021, the press event raised all kinds of questions. Foremost among them was whether this was just another media mirage cooked up by an untrustworthy president.

The intriguing results Slaoui was referring to came out three days later in a press release from Moderna. The company mentioned the virus challenge trials conducted by Ralph Baric's team in North Carolina, noting that the vaccinated mice had all been protected from infection.

The headline news was the neutralizing antibodies in the blood serum from eight volunteers. They were at or above those seen in patients who had recovered from COVID-19. When Barney Graham saw the results, he was floored. Normally, the results are reported in terms of the highest level of dilution of a vaccine that can achieve a 50 percent reduction in virus infection. In this case, the vaccine was achieving 100 percent neutralization at surprisingly high dilutions.

Moderna's stock price closed that afternoon at eighty dollars per share, more than triple the price in early January. The company's market value was estimated at thirty billion dollars. "The market reacted with reckless abandon," *Mad Money* host Jim Cramer enthused. "If anybody can develop a vaccine in record time, it is Moderna . . . this could be one of the shortest recessions in history." Stéphane Bancel sold off nearly 17,000 shares, netting $1.2 million. His chief medical officer, Tal Zaks, raked in nearly $10 million from his stock sales.

While the nation was feeling hopeful and investors were feeling giddy, the scientific community pounced. "[Moderna] revealed very little information — and most of what it did disclose were words, not data," Helen Branswell, the grande dame of infectious diseases reporting, wrote at Stat, a medical-news site. "My guess is that their numbers are marginal or they would say more," an expert griped to Branswell.

Adrian Hill, the director of Oxford's Jenner Institute, added to the Moderna pile-on. In an interview with CNN's senior medical correspondent Elizabeth Cohen, he lashed out at Oxford's competitors. He considered mRNA vaccines "noise from the new boys." The NIH-Moderna vaccine, he said, was "weird and wonderful." Cohen asked him what exactly he meant by "wonderful."

"I was being sarcastic," Hill said.

When Moncef Slaoui took the helm of Operation Warp Speed, it was less than operational and it wasn't moving at warp speed. Over the previous month, the teams at NIH and BARDA that were already coming together on the effort had been struggling with a central challenge: the leading vaccine makers weren't necessarily eager to be a part of a big bureaucratic program like Warp Speed. One major problem was that the government was no longer handing out bags of cash so the companies could go off and run their own phase 3 clinical trials. No, now companies had to participate in Peter Marks's elaborate framework — the single master protocol that would pit the vaccines of these fiercely competitive organizations against one another in a horse race. The results would be unequivocal and the winner would take all — or at least have bragging rights.

Moderna's leaders had been caught off guard when they learned they were going to be asked to participate in a megatrial like this. That was not part of the plan back in mid-April, when BARDA, under an agreement that predated Warp Speed, plunked down $483 million for their trials. Nor was Barney Graham enamored with this master-protocol idea. To think that Moderna in the face of a competitor like Pfizer would tether the start date for its clinical trials to the laggards in the race, like Sanofi and Novavax, was outrageous! If Pfizer's mRNA vaccine made it to the finish line early, why would anyone volunteer to be jabbed with an unproven competitor? And what if there was a safety problem with one of the vaccines in the master protocol? Because the researchers and volunteers were supposed to be blinded to who received which company's vaccine, the entire trial would have to come to a halt.

Four hundred and eighty-three million dollars was a lot of money for any company to turn down, but the upside to going it alone was alluring, and investors were bullish. If Moderna was cold toward the master-protocol concept, other companies might be as well. On April 28, the director of the Vaccine Research Center, John Mascola, sent an apologetic e-mail to Moderna CEO Stéphane Bancel and chief medical officer Tal Zaks. "I know that there has been some inconsistency in communications and expectations, so I ask for your forbearance there — during this public health emergency," he wrote. Mascola explained that the view inside government was that the vaccine trials being funded now would need to continue to be closely coordinated by the federal leaders, but, he said, the different agencies involved were working to agree on a more flexible model, one in which the companies would maintain greater control of how their data was used.

In the meantime, Francis Collins was trying to convince Pfizer to take part in the government's master protocol. Collins had a collegial relationship with Mikael Dolsten, Pfizer's chief scientific officer. In March, Dolsten's wife, Catarina, had developed COVID-19, and she became so ill that she had to be admitted to Mount Sinai Hospital in New York City. Dolsten had sought advice from Collins, who put him in touch with one of the NIH's top infectious diseases doctors. After Catarina recovered, Dolsten felt surer than ever that a vaccine was needed and he'd told Collins as much. On May 3, Collins e-mailed Dolsten and said that a master protocol with Moderna, Sanofi, and Johnson and Johnson was set to launch July 1. "Might Pfizer also join?" he wrote.

Dolsten told Collins that this was going to be a nonstarter. Pfizer and its partner BioNTech had whittled twenty vaccine candidates down to four on April 12. Pfizer was planning to start testing all four of those in human volunteers, and it would evaluate those data on the fly to make a final selection. It was a miniature Operation Warp Speed, the kind of bold, costly program that only a powerhouse like Pfizer could afford. "By July we will hopefully be far into a potential pivotal phase 2/3 study . . . we don't want to slow down to wait," Dolsten wrote back. "We have a possible chance to be crossing the finish line way before the timelines shared by JnJ and Sanofi." Dolsten ended with what one might call a "soft no," suggesting that Pfizer might agree to add "one of our several vaccine formats that is less advanced."

Collins had to concede the company was better off going it alone. "We would certainly not want to do anything to slow you down!" he replied.

However, a deal to bring ChAdOx1 into the Operation Warp Speed fold was imminent. This was the viral vector vaccine from the Oxford team now being developed by AstraZeneca. On May 13, the Oxford team had shared with the scientific community a prepublication draft of a paper containing the first results from vaccinated rhesus macaques challenged with live virus. The results of these experiments, which had been conducted at NIH's Rocky Mountain Laboratories, outside of Missoula, Montana, weren't easy to parse.

While the lung washes from monkeys vaccinated with ChAdOx1 were almost entirely free of virus, the vaccine made no difference in the amount of virus measured in nasal-swab samples. That meant the vaccine was protecting against severe disease but not preventing infection. In that case, the virus could still spread even among vaccinated individuals. In addition, ChAdOx1 wasn't generating the hoped-for levels of neutralizing antibodies. The serum from these monkeys was losing its efficacy with a fortyfold dilution.

There were also some reservations at NIH about the fact that the gene sequence delivered to cells by ChAdOx1 did not make use of Barney Graham's innovations. Unlike Johnson and Johnson's vaccine, the spike proteins that this vaccine generated would be unstabilized and likely to flip to a postfusion state, which would produce fewer neutralizing antibodies. Nevertheless, the Secretary's Science Advisory Council voted unanimously on May 14 to move forward with finalizing the contract.

With AstraZeneca entering the picture alongside Moderna, the NIH's John Mascola was advocating for a sort of compromise on the master protocol, what he and his collaborators called a "harmonized protocol." The idea was that companies could start their own trials at any time, but they would follow a generalized blueprint negotiated with Operation Warp Speed. They would then tap into the NIH's networks of clinical trial sites, mostly academic hospitals, that Tony Fauci had fostered for HIV studies. The trials would use the same immune-monitoring labs across the board for assays and rely on some of the same statistical analyses. The vaccines could still be roughly compared to one another after the fact, but independent trials meant the companies could conceivably all claim victory on their own terms, which was the way they liked it. Having comparable data from multiple trials would still allow the government to come up with a metric, known as correlates of immunity, that would make possible the approval of coronavirus vaccines in the future without a full-fledged clinical study, as was done with annual flu shots.

Peter Marks at the FDA was worried about what would happen if, under the harmonized protocol, multiple trials were competing for a limited pool of volunteers or if a mediocre vaccine was approved while the other trials were still running. But he conceded that there were some advantages to the new protocol idea — namely, actually being able to convince companies to participate. He was still insistent, however, that the Operation Warp Speed portfolio be selected, at least in part, based on the animal trials he had set up. This would give the government a truly comparative picture of all the vaccines it was funding to ensure it was making the best investments. Marks's model for the animal trials could also ensure enough validated animal data to justify deploying the vaccines before the clinical trials came to an end.

Arriving at the Humphrey Building on the afternoon of May 20, Marks had a bad feeling about the five p.m. board meeting, the first with Moncef Slaoui in charge. A couple of days before, he had sent Slaoui slide decks along with detailed spreadsheets and plans he had laboriously compiled for the animal trials. Weeks of work. Weekends too. What had he gotten in return? Crickets.

In Azar's conference room, Marks found himself a seat at the table. Practically the whole crew was there: Jared Kushner, Adam Boehler, Tony Fauci, Francis Collins, Gustave Perna, Bob Kadlec, Paul Mango. Deborah Birx couldn't attend, but she would be briefed about the meeting later.

Slaoui announced to the group that there was no need for those extra animal studies. Well, he didn't say that exactly. He just didn't bother to mention those studies. He had selected his own portfolio. He foresaw five or six vaccine candidates, down from the fourteen Marks had his eye on. Four of those candidates, Slaoui said, could credibly deliver a vaccine by the end of the year. He listed the mRNA-based vaccine from Moderna, the protein-based vaccine from Sanofi, and the viral vector vaccines from AstraZeneca and Johnson and Johnson. He thought that Novavax might have a chance down the line with its protein-based vaccine and noted that Pfizer was independently working on its own mRNA vaccine, one he felt that Warp Speed should be monitoring. After weeks of Marks trying to hone his persnickety process for picking the perfect portfolio, it was stunning for Azar and Kadlec to witness Slaoui act with such decisiveness.

Peter Marks had agreed to the harmonized approach proposed by NIH, but now he felt Slaoui was destroying the central feature of his original Project Warp Speed. The only element still in place was the name. Early the next morning, Marks sent an e-mail to Kadlec, asking if they could talk. "I have reached a decision regarding my further involvement in Warp Speed," he wrote. On the phone at seven a.m. that Friday, Marks told Kadlec he felt he no longer had much of a voice in Operation Warp Speed and that his romantic notion of a team effort had been blown up by Slaoui. Kadlec listened sympathetically as Marks spouted off. He didn't interrupt him or challenge him or try to clamp down on his diatribe or talk about his own feelings about the paradigm shift. He just listened.

Eventually, Marks came to a conclusion: There was no way he could have been a part Operation Warp Speed anyway, not if he wanted to keep his job at the FDA. It was a conflict of interest for him to judge the same vaccines he'd had a hand in selecting. Come fall, when these vaccines emerged from the pipeline, his team at the Center for Biologics Evaluation and Research was going to need his leadership more than ever.

Kadlec thought that made perfect sense, and he let Azar know what had happened. Azar told Marks to wait until the two of them spoke that afternoon before he officially resigned. Like Kadlec, he knew how important it was to have someone inside the FDA who was sympathetic to the Operation Warp Speed mission. The last thing he needed was another media controversy interfering with his broader policy goals or threatening his job. He asked Marks to reconsider his resignation, even though he knew it was a done deal. "I'm so grateful for what you've done, Peter," Azar effused. "I'm glad you'll be at CBER."

Matt Hepburn was leaving the Pentagon that afternoon and walking to his car when Peter Marks called. "We're going to need you as the vaccine lead," Marks told him. As one of the Wolverines, Hepburn was close to Bob Kadlec, and he was originally just going to be a foot soldier working underneath Marks on the Warp Speed vaccine team. Now he would essentially be Slaoui's deputy, overseeing the entire portfolio.

After the call with Marks, Hepburn was at a loss. He stood in the enormous Pentagon parking lot taking it all in, thinking about the weighty responsibility. A few days later, Marks passed on to Hepburn a plastic bag filled with those Warp Speed buttons he had designed. They were a lavender color, lighter than on the official *Star Trek* uniforms. "The purple didn't come out exactly right," Marks said. He still wanted Hepburn to have them.

That evening, May 21, Alex Azar called Jared Kushner from the kitchen of his home in Maryland to tell him that they had closed the deal with

AstraZeneca. "The whole world wants this vaccine, and I've just nailed three hundred million doses with the U.S. at the front," he boasted. If that vaccine proved efficacious with a single dose, more than 90 percent of the U.S. population would be covered. Kushner was elated. He was with a crowd that included Hope Hicks, the White House communications adviser; Dan Scavino, the deputy chief of staff; and Trump. He put Azar on speakerphone to talk directly to the president. Azar again explained how great the AstraZeneca vaccine was. "It's going to be announced at four a.m. tomorrow when the British stock markets open," Azar said.

"British?" Trump said. "Is AstraZeneca a British company?"

"Yes, it is, but it has significant U.S. operations, in Delaware in particular," Azar replied.

"A British company! This is terrible! Terrible. I'm gonna get killed. Oh, Boris Johnson will have a field day with that." Trump sighed. "Well, I said there's no ego here so I guess that's okay . . . but I don't want to have any press on this!" After Kushner hung up with the dejected health secretary, Hope Hicks tried to explain to Trump that he should think about it as a win for America, not a loss for himself. The next day, a Friday, a press release went out from HHS about the deal, which Azar called "a major milestone" for the Trump administration's "multi-faceted strategy for safely reopening our country and bringing life back to normal."

At noon the following day, Saturday, May 23, John Mascola spoke to AstraZeneca's head of vaccines, Menelas Pangalos, to learn about the state of the company's vaccine trials around the world. Mascola could sense some unease on Pangalos's part over the bureaucracy he was being asked to submit to. Afterward, Mascola told Francis Collins that they'd have to proceed delicately to ensure there were no hang-ups with AstraZeneca's willingness to make use of the academic clinical trial sites to recruit and follow volunteers. Most companies were going to prefer the speed and simplicity of working with commercial research sites that they had preexisting relationships with. Collins told Mascola he was having his own call with Pangalos the next day and would assure him that NIH "is prepared to move with breathtaking speed."

On the call, AstraZeneca's Pangalos, in turn, communicated the company's plans to move forward with a single dose of its vaccine in its phase 3 trial in the United States. But the company intended to wait for the data from the United Kingdom trials over the next two to four weeks. "If that indicates that a second dose is needed, [AstraZeneca] would want to be able to modify the U.S. protocol," Collins wrote afterward to Mascola.

The launch of Operation Warp Speed coincided with the end of any coordinated pandemic response by the White House. Jared Kushner believed that the best strategy for improving his father-in-law's chances in the upcoming election was to make Trump into a "cheerleader" playing to the "psychology of the market." No attempt would be made to balance the nation's economic health with a calibrated public health strategy. The White House would leave it to individual states to choose their own path through the crisis and deal with the blowback that came with their individual public health policies. "The federal government should not own the rules," Kushner said that month. "It's got to be up to the governors."

In his April 27 press conference on reopening, Texas governor Greg Abbott told cities they could no longer enforce mask restrictions. Nail salons and barbershops would be back in business on May 8. Texans could return to their bowling alleys, bingo halls, and skating rinks on May 18. Utah governor Gary Herbert was allowing the reopening of gyms and indoor dining. Missouri governor Mike Parson eliminated restrictions on social gatherings as long as people could maintain six feet of distance from one another. "It's supposed to be one of the biggest Memorial Day weekends we've seen at the lake here in years," said the manager of a resort at Missouri's Lake of the Ozarks. Backwater Jack's bar and grill at the Lake of the Ozarks started gearing up for its third annual "Zero Ducks Given Pool Party."

Within weeks, a lull in coronavirus cases ended and hot spots began popping up in some of these same states. An outbreak at a meatpacking plant in the Texas Panhandle gave Moore County the state's worst infection rate — 2.5 cases for every 100 residents. Down in Houston, an executive at the Methodist Hospital system told staff members to delay performing elective surgeries in order to prepare for a crush of COVID-19 patients.

Along with "not owning the rules," scapegoating continued to be another essential part of the White House coronavirus strategy. On Friday, May 29, Trump emerged from the Oval Office to deliver a fiery afternoon press conference in the Rose Garden. China had reported no new confirmed cases on the mainland for the first time since January, but numbers were continuing to grow within the U.S. borders and in other countries, including India, Mexico, and Brazil. "China's cover-up of the Wuhan virus allowed the disease to spread all over the world, instigating a global pandemic that has cost more than a hundred thousand American lives and over a million lives worldwide," Trump said. In less than two months, the country had blown past Fauci's estimate of 60,000 deaths by the fall.

Setting Trump's rhetoric aside, it was true that the origin story of the coronavirus as a natural spillover event remained murky. How and when did the virus make the leap from bats to humans? Early on, one group of Chinese researchers posited that the virus could have passed through Chinese cobras, which were sold at the Huanan Seafood Market. Others suggested it could have been spread via the illicit trade of the scale-covered mammals known as pangolins, used in traditional Chinese medicine. Although some early infections were linked to the wet market, even George Gao of the Chinese CDC couldn't find evidence that the virus had come from any animals sold there. The heavily trafficked market seemed to have been a center for amplifying and spreading the virus around Christmas and New Year's, but there was genetic evidence pointing to at least a few cases of COVID-19 circulating as early as October 2019.

The Chinese were not keen to shine further light on the matter. Cliff Lane, the top infectious diseases doctor at NIH, took part in a WHO mission to China in February 2020, but he was told the participants would not be hunting for the animal source or reviewing China's missteps in the early days of the outbreak. Many in government and academia were, at least privately, leaning toward the theory that it could have been an accidental release from one of China's virus laboratories. It wouldn't be the first time this had happened. In April 2004, two researchers studying the first SARS virus at the Beijing Institute of Virology were infected in separate incidents and they spread the virus to seven others. Hundreds were quarantined and one person died.

Zhengli Shi — the "Bat Woman" at the Wuhan Institute of Virology — had questioned in January 2020 whether the new coronavirus infections had been a repeat of this scenario. Not only did Shi's institute keep live bats on its premises, some of Shi's coronavirus research had been conducted under the lax conditions of biosafety level 2 laboratory, roughly equivalent to a dentist's office. She'd even asked herself, "Could they have come from our lab?" she said in a March 2020 interview with *Scientific American*. The bats that harbored coronaviruses lived in the southern part of the country, where she did her fieldwork. "I had never expected this kind of thing to happen in Wuhan, in central China," she said.

Shi ruled out a lab escape, but her team had sequenced the closest known relative of SARS-CoV-2, RaTG13, which she had collected from bat droppings at the entrance to an old copper mine in Yunnan Province in 2013. Several aspects of her research program soon began to raise suspicion. She had identified eight other SARS-virus-like coronavirus sequences in the samples from the mineshaft over several years, but she had never shared the sequence data with the scientific community as is standard practice. She had also omitted what had prompted the group to sample that particular mineshaft: Six miners there had fallen ill with a SARS-like illness in April 2012. Three of them died.

For the China hawks in the White House and others who wanted to minimize the administration's failings, the most damning evidence came from a U.S. intelligence report claiming that, in November 2019, three individuals from the Wuhan lab had become sick enough they had to be hospitalized. That report had yet to become public, but during Trump's speech in late May, he laid the blame for all the pandemic problems on the World Health Organization for treating China with kid gloves and repeatedly delaying announcements on the state of the threat the world was facing. "China has total control over the World Health Organization," he said. "We have detailed the reforms that it must make . . . but they have refused to act. Because they have failed to make the requested and greatly needed reforms, we will be today terminating our relationship with the World Health Organization."

It seemed as if the forty-fifth president of the United States was preparing the nation for war.

17

BATTLE RHYTHM

P eter Marks's *Star Trek*–inspired logo was supplanted by a more polished emblem for Operation Warp Speed, a hybrid of the official seals of the Departments of Defense and Health and Human Services with a coronavirus capsule at the center and radial streaks of stars zooming past. The Vaccine Operations Center was established in a wing on the second floor of the Humphrey Building. General Gus Perna, Moncef Slaoui, and other members of the leadership team had fancier digs up on the seventh floor. Uniformed army personnel marching on the brown carpets in their combat boots soon became a familiar sight. There were navy and air force folks in the mix as well. Over the summer, dozens of them would squeeze into the stuffy conference room on the seventh floor, cooled by two fans. Face masks were nowhere to be seen. "The military invasion of HHS," griped certain public health officials in the building.

The potential culture clash was one of the things Bob Kadlec had warned Perna of during their first one-on-one call on May 14. Kadlec knew Perna through a mutual friend, and he held the general in high esteem. He asked Perna if he had ever read the Holloway Report, a review of Operation Eagle Claw, the disastrous Special Operations mission to free fifty-two hostages in the U.S. embassy in Tehran. The botched operation left eight servicemen dead. Perna hadn't read it. The failure, Kadlec said, came down to political interference and a sloppy command-and-control. He drew out the analogy and warned the general that Perna would inevitably butt heads with the scientific fiefdoms inside HHS—namely, the CDC and the NIH—that were controlled by powerful personalities. As the leader of the operation, Perna faced a complex dilemma: the mission depended on the science, but you couldn't let the scientists run the mission.

The topic of that first call soon gave way to a more urgent matter. Kadlec, who had previously served on the Senate Intelligence Committee, was deeply concerned about the security risks posed by Operation Warp Speed. Leaks of privileged information to the press were the least of it. Twice under Kadlec's watch, individuals from potentially hostile countries had gained access to BARDA's technical databases, which held information on medical countermeasures. More broadly, in March 2020, the HHS experienced the most significant cyberattack in the agency's history. Its networks, which have more than a million ports, or entry points, where users can connect, is larger than that of some small countries. The denial-of-service attack taxed the agency's servers to such a significant degree that its chief information officer, Jose Arrieta, had to shut them down for ten minutes in order to foil the bots.

Perna raised the issue up the chain of command. It was like a snap of the fingers. Brett Goldstein was upstairs in his home in DC getting one of his three kids dressed when he heard what he referred to as the "Oh, shit" ring—a long, insistent trill. Goldstein, a shaggy middle-aged dude who looked more Silicon Valley than Pentagon, ran a nimble group of coders called the Defense Digital Service. When the Pentagon put social-distancing measures in place, he had brought home an overwhelming amount of gadgetry with which to communicate and test out software. The "Oh, shit" ring meant that the military brass was trying to contact him via its communications division, called the Cables Branch, forcing Goldstein to scramble frantically around the house to find the cell phone that was going off before the next phone in his communications arsenal buzzed. Finally, with a stack of phones in his hand, Goldstein answered the call on his Apple Watch. It was the deputy secretary of defense, David Norquist, telling Goldstein to ready his team.

The Humphrey Building had a SCIF, where the most sensitive Operation Warp Speed discussions took place. On the morning of Tuesday, June 2, Kadlec and Jose Arrieta, the HHS chief information officer, had arranged a videoconference that included Goldstein and Anne Neuberger, the head of the National Security Agency's Cybersecurity Directorate. Neuberger had been working closely with Arrieta since the March cyberattack, deploying one of the agency's secretive tools, called Overwatch, to block suspicious web traffic heading to HHS from foreign networks. Now they needed to game out a plan for protecting Operation Warp Speed. Compared to the DoD's fortresslike network architecture, HHS was a bale of straw.

The day before, Arrieta and Kadlec had spoken to the NIH team at the Vaccine Research Center and learned that handling clinical trial data would involve no fewer than twenty different software applications. The NIH was an organization designed to share information, not protect it from nefarious state actors. This was unfamiliar terrain for them. Inside the SCIF, Kadlec painted a scenario where China or Russia might potentially steal or alter clinical trial data to harm vaccine confidence. The FBI was already investigating a pair of Chinese hackers who had been probing for vulnerabilities in the networks of Moderna and Novavax. "This is war," Kadlec said.

For the day's meeting, Arrieta had pulled in one of the top cops in the HHS's inspector general's office, which was charged with making sure the tech team didn't violate any patient-privacy laws. The two men described to Neuberger their proposed strategy for Warp Speed's information flow. "Let's limit our surface area" is how Arrieta put it. Rather than try to defend HHS's entire porous network, they would create a private network to house all the information connected with Warp Speed's clinical trials.

"We can run Overwatch on that," Neuberger said.

Kadlec went one step farther. HHS would buy new secure phones and laptops for all Warp Speed participants to keep everything siloed. If people were calling into a meeting taking place in the SCIF, Neuberger pointed out, they would not be able to speak from a remote, unsecured location. Instead, they would have to use Signal, an encrypted text-messaging application, to send a message to a person in the room. Neuberger also suggested the group use Slack to share data files, as that messaging software could be readily secured as well.

Brigadier general Mike McCurry, who would be supervising the overall security effort for Warp Speed, proposed sending two members of his team to the NIH campus in Bethesda. They'd snoop around the Vaccine Research Center to hunt for any physical vulnerabilities. Vaccine makers would also get a security assessment. In the coming weeks, the security team would establish battle rhythm, meeting once or twice a week to receive global security briefings from the NSA and review any foreign investments, particularly Chinese investments, coming into U.S. companies involved in Warp Speed. These measures would succeed in protecting the operation in the critical months to come.

Two days later, an Operation Warp Speed consultant named Carlo de Notaristefani arrived at the blocky white buildings housing Emergent BioSolutions, a factory for hire in Baltimore's Bayview neighborhood. Another contract manufacturer, Lonza, would be focused solely on Moderna's mRNA vaccine at a facility in New Hampshire. Emergent, by contrast, was working with three different Warp Speed candidates using more traditional biological manufacturing methods that employed cell and virus cultures.

De Notaristefani knew that Emergent had been dinged for quality-control issues during an FDA inspection in April and that the company was going to be operating at a scale far beyond anything it had ever attempted. To reach the target of three hundred million doses of an approved coronavirus vaccine by January 2021, Operation Warp Speed needed to aim for two or three times as many in case one or more failed. Achieving that goal was going to take all the manufacturing might the United States could muster. At one time, the country had a network of state-owned vaccine plants, but flagging investment had left them in a shambles. Many had closed down.

In 1998, a retired navy admiral bought the very last one — the Michigan Biologic Products Institute, a decrepit facility near the airport in Lansing that had been hemorrhaging millions of dollars per year. It was the country's only licensed manufacturer of anthrax and rabies vaccines and had been on the market for two years with no takers — at least until the Clinton administration began talking about going to war with Iraq. Then the Pentagon announced it planned to invest fifteen million dollars to double the plant's capacity in order to vaccinate troops.

The company running the plant became Emergent BioSolutions, and over the next twenty-two years, it grew fat on the government's dime, selling products to the Pentagon and the Strategic National Stockpile. In the aftermath of the 2009 swine flu pandemic, BARDA recognized the flagging state of the U.S. vaccine armamentarium and doled out contracts to three factories, among them Emergent's Bayview facility, which received $163 million. But as the memory of that scare faded, BARDA's investment trailed off. Emergent had filled small orders for Ebola and Zika vaccine candidates, but that facility was mostly sitting empty, losing money. Same old story.

When Operation Warp Speed started coming together in late April, BARDA realized it needed to lock down capacity and it shelled out $600 million to take over the whole Emergent facility. Now de Notaristefani was looking around and kicking the tires, so to speak.

For the past several months, the Emergent team had been setting up Area 2, the manufacturing space for Johnson and Johnson's viral vector vaccine. Specially engineered cell cultures would be incubated inside a two-thousand-liter bag kept at a constant temperature. Once there was a sufficient number of these cells, they would be infected with J and J's adenovirus vector, which would multiply in these cells and serve as the raw ingredient for the company's vaccine. Growing viruses at that scale can be as fussy as making fine wine. Even if you use the same "master virus seed" to start each batch, you may not end up with the same result every time. Problems can occur anywhere, from growing the virus to filtering out the parts of the animal cells it was growing in. That's why the team was trying to reproduce the exact setup that Johnson and Johnson had constructed in the Netherlands for the large-scale production of its successful Ebola vaccine, which used the same vector (meaning the same virus). Ideally, the team in the Netherlands would have flown over to help with the setup and the months of troubleshooting, but COVID-19 had forced everything to be done over videoconferencing. In some cases, Emergent had to cut new doorways in its Bayview facility so that the layout exactly matched that in the Netherlands.

The AstraZeneca vaccine would be installed in the adjacent suite, Area 1. Emergent would have to be scrupulous to prevent cross-contamination, especially as both vaccines used hardy adenoviruses as vectors. For instance, employees leaving Area 1 would have to shower before entering Area 2 (and vice versa). Novavax's protein-based vaccine candidate was located in Area 4, which could be used to grow vaccines in batches of up to fifty liters only. Depending on how the vaccine trials and scale-up progressed, the front-runner could be installed in Area 3, the larger "pandemic suite" that had been funded by the government. If past experience was any guide, one or more vaccines would end up failing in clinical trials, opening up space for the winner.

As de Notaristefani toured the plant, he could see the vast effort that would be required for the scale-up. In a report he prepared for the Warp Speed team, he identified what he called "key risks" at the facility. The quality-control problems, he warned, "will require significant effort" and "will have to be monitored closely." The company's staffing plans, too, seemed "inadequate to enable the company to manufacture at the required rate."

Warp Speed would certainly need to have a person in-plant at Emergent and other facilities who could sit in on meetings and report back to the Humphrey Building. De Notaristefani also felt that success required that Emergent not spread its resources too thin. He believed that the Novavax vaccine should be kicked out of the facility after Emergent completed manufacturing the several hundred doses needed for the phase 1 trials. Liter for liter, a viral vector vaccine like AstraZeneca's gave you a lot more doses than a protein-based vaccine like Novavax's.

The Warp Speed team was figuring out how to do things that had never been done before, or at least had never been done during a pandemic. As the team discovered, vaccine-making capacity was just one resource that was at a premium on the open market. But Operation Warp Speed didn't have to rely on the open market. It could make use of the 1950 Defense Production Act.

Most contracts the military signs, such as those for helmets and missile parts, come with a priority and allocation rating that puts them ahead of commercial orders and ranks them with respect to one another. In 2012, President Obama delegated HHS the authority to rate contracts for "health resources," but the agency had rarely, if ever, employed that authority. President Trump had FEMA and the DoD invoke the Defense

Production Act to obtain masks and ventilators during the supply-chain crisis in April. Other clauses in the Defense Production Act could force a company to build out manufacturing capacity or even force two competitors to work together, as the government had done during World War II, when automakers built tanks for the war effort.

On June 9, retired lieutenant general Paul Ostrowski, General Perna's deputy, walked over with Perna for a meeting in Alex Azar's conference room. Ostrowski had spent the previous months helping obtain medical supplies for FEMA and would now be in charge of turning Perna's orders into action. He was carrying a brief about the ways in which he foresaw the team using the Defense Production Act.

As Bob Kadlec, Alex Azar, Paul Mango, and the top HHS lawyer Bob Charrow sat around the conference table, Ostrowski spoke about how Operation Warp Speed might have to order a clinical research company to run coronavirus vaccine trials or direct other companies to distribute a product. The Defense Production Act authority had to be used thoughtfully, however. Forcing a company to make a vaccine might lead it to kick a lifesaving cancer-drug candidate, for instance, to the curb. If the government was helping a multinational company buy a hard-to-obtain filter or some other part, it had to be certain it was being used exclusively for the benefit of Americans. With a complicated program like Operation Warp Speed, you needed medical-supply-chain experts to game it all out.

One of the issues that came up at the meeting was where the authority to evaluate Defense Production Act requests would rest. Charrow had previously questioned whether Kadlec's suggestion — to delegate the authority to him — would be legal, and Azar was hesitant to deploy it on his own. The group decided that such sensitive decisions should be made during the weekly or biweekly board meetings, which would bring in both Azar and Defense Secretary Mark Esper or his deputy David Norquist and would be sure to have the full backing of Jared Kushner.

As many a White House administration has learned the hard way, going to war is not cheap. Engaging all the might of the military at this kind of scale was going take some serious cash. For weeks, Moncef Slaoui and General Perna worked with the finance team at HHS to come up with a budget for Operation Warp Speed. It was one thing to say that money would be no object and quite another to see the figure in bold print on a funding proposal: thirty to forty billion dollars for vaccines and drugs. Even with Kushner behind it, everyone knew that Russ Vought was going to have a conniption fit. He was the tightwad with the devil horns over at the White House budget office, and he knew that HHS was flush with funding from three coronavirus supplement bills.

Vought's office sent back a menu of crafty options for moving around the pots of money. First, HHS could yank a billion dollars or so from the Strategic National Stockpile that he loathed. The agency could also take about ten billion dollars from a massive fund that congressional Democrats had directed to support hospitals and local health clinics. When HHS tried to tell Vought that it would need to notify Congress about such a move, he told the agency to hush up. His office issued a legal opinion that there would be no problem with the transfer, and HHS went along with it. Last but not least, Vought suggested a garage sale: Warp Speed could raise at least a few hundred million dollars selling all those Hanes masks that Kadlec had wasted money on. "Half of them had already gone out the door," said one incredulous HHS official. The message from the White House was clear: *This is your payback.*

Operation Warp Speed continued to be a useful foil in diverting attention from the growing coronavirus death toll, which was rising due to the administration's rejection of masking and its refusal to organize a federally coordinated public health response. One of the Faustian bargains Alex Azar had made in order to survive was allowing Michael Caputo's propaganda shop to take full control of the messaging coming out of HHS. Caputo was working on an effort to further the narrative that vaccines were on the horizon and the worst of the coronavirus was behind us. Keep your eye on the vaccines, in other words.

Caputo saw celebrities as a central pillar of that effort. On April 30, a member of his communications team listened in on a twenty-seven-person Zoom call organized by Kim Kardashian that included the singer Ariana Grande, NBA player Chris Paul, actors Ashton Kutcher and Mila Kunis, and Tony Fauci. One particular moment that stuck with Caputo was when Kunis, Kutcher's wife, asked if Fauci still ordered takeout. Fauci said he still ordered from his favorite joint in his neighborhood, though he always wore a mask and gloves when he picked it up. "Do you eat out of the box?" Kunis asked incredulously.

Fauci said he did.

"See!" Kutcher exclaimed. "That's it. I'm ordering a pizza right now."

The White House budget office agreed to transfer three hundred million dollars from the CDC's budget to Caputo's office so that he could launch an all-out effort to "defeat despair about COVID-19." He hoped to use interviews between celebrities and the administration's top doctors to promote the vaccine and reassure the public about its safety.

Although the propagandist in chief had some of the nation's leading health experts at his disposal, he chose to bring in an outside consultant, a Canadian professor named Paul Alexander who had been a guest on the conservative radio program Caputo used to host in Buffalo. Both Trump and Caputo felt that, across the organization, career scientists were not there to inform the public or guide policy but to serve their political masters — or at least not work against them. With suspected leakers like Rick Bright and Joe Grogan gone and no longer able to make Trump look bad, Caputo began to see the CDC career staff as his next enemy.

One press release the CDC had planned to issue on May 28, for instance, announced the agency's first data on COVID-19 hospitalizations broken down by race and ethnicity. According to a draft of the press release, Black and American Indian populations had hospitalization rates more than four times higher than whites. Alexander observed that the press release was "very accurate." He added, however, that "in this election cycle that is the kind of statement coming from the CDC that the media and Democrat antagonists will use against the president." Alexander said that Trump's enemies were already accusing Trump "directly of the deaths in the African American community from COVID" and he felt that the CDC's press release needed to be contextualized with information on how Democrats had, for decades, neglected people of color. The CDC's press release never went out.

These were challenging times for Alex Azar. During a once-in-a-hundred-years pandemic that had infected two million Americans, the functioning of our nation's health department hinged not so much on the laws and regulations on the books but on maintaining trust between those working inside the various agencies and the political appointees leading them.

Azar wanted to reassert his authority on a number of matters, but he was not a military commander, and his scientists and department heads were not his soldiers and generals. If he could not create at least the appearance of unity within HHS, then what chance was there for the American public to trust the vaccines? Vaccines that he knew were about to be deployed on a national scale in the first phase 3 clinical trials.

18

MASTERS OF THE PROTOCOL

Researchers who run human clinical trials often say that mice lie and monkeys exaggerate, but everyone watching the progress of the Moderna vaccine was eager to hear the reports from the animal world. Barney Graham, at the NIH, knew by now that the mRNA vaccine could prevent a coronavirus infection in mice and that it could generate neutralizing antibodies in monkeys, but could it prevent an infection in monkeys? Those were the data he wanted to see before Moderna's phase 3 trial began, when thirty thousand people would be taking part in the largest mRNA vaccine experiment in history. The results of that landmark study would determine how far away humanity was from an end to the pandemic.

The first samples from the monkey studies would be a sneak preview. Twenty-four rhesus macaques had received two shots of either mRNA-1273 or a placebo at the site of an old farm the NIH had purchased years ago to house its large research animals. At the end of May, those monkeys were placed in steel crates and trucked over to Bioqual, the contract laboratory in Rockville, Maryland. There, they were given several weeks to develop antibodies before facing the final test. On June 8, the Bioqual team put on their biosafety suits, headed down to the secure laboratory where the monkeys were kept, and squirted live coronavirus up their noses. Over the next seven days, samples were taken from the monkeys' noses and lungs, and blood was drawn. The monkeys were subsequently euthanized so that their lung tissue could be examined in detail under a microscope. Kizzmekia Corbett, Graham, and the rest of the team ana-lyzed samples from the monkeys throughout the experiment, measur-ing the neutralizing antibody levels and comparing them with those in people who had recovered from COVID-19.

To quantify the amount of viral RNA in both the vaccinated and un-vaccinated monkeys, the research team tested samples from nasal swabs and lung-wash specimens. They found that two days after the virus chal-lenge, all but two of the vaccinated monkeys had beaten back the virus in the lungs, hinting that the vaccine would help prevent severe COVID-19. The unvaccinated monkeys, meanwhile, were still fighting it. The vac-cine provided a further benefit: in the monkeys that had received the highest dose, the virus was all but eliminated in the nose by day two, sig-nificantly decreasing the likelihood of the virus spreading among vacci-nated individuals.

The results were as good as any Graham had ever seen. Corbett, in drafting their paper on the study, took a measured stab at the Oxford group, pointing out that the ChAdOx1 vaccine, by contrast, had "pro-vided no evidence of a reduction of viral replication in nasal tissue, rais-ing questions as to whether [it] could affect virus transmission." It was tempting to compare results across the monkey studies, but the Oxford group had actually infected their monkeys with more virus particles.

Based on the Moderna results, it seemed likely that antibodies — rather than the more complex T cell response — were the primary mech-anism protecting the monkeys. This didn't guarantee that antibodies alone would protect a person (or a monkey), but it was highly likely. Graham had also seen the results from hundreds of volunteers partici-pating in the phase 2 study, confirming that both the low and high doses produced an adequate immune response with no serious side effects. Taken together, these latest pieces of data were the critical hurdles that needed to be leaped before Graham and the FDA would agree to put more people at risk in Operation Warp Speed's large-scale phase 3 hu-man trials.

John Mascola, the director of the Vaccine Research Center, had spent fif-teen years in the navy, including one as a medical officer for a destroyer squadron accompanying the battleship USS *New Jersey* on a tour of the

Western Pacific. From the east-facing window of his office on the NIH campus in Bethesda, he could make out the art deco–style tower of the Walter Reed National Military Medical Center, where he had trained in virology under Robert Redfield and Deborah Birx. He had a low-key demeanor and a military man's sense of mission. But for the job in front of him — planning those upcoming phase 3 clinical trials — he was effectively sharing power with a much larger personality, Larry Corey of the Fred Hutchinson Cancer Research Center in Seattle.

With wavy white hair and glasses, Corey was a distinguished virologist and doctor, and he didn't let anyone forget that fact. An academic through and through, Corey unselfconsciously referred to professors at his former department at the University of Washington as "my disciples." Before the pandemic, Corey came to DC at least once a month to meet up with his old friend Tony Fauci, and the two men would discuss their HIV clinical trials over sea bass and scallops in the back of Matisse, a French-Mediterranean restaurant on Wisconsin Avenue.

More recently, he would call in from his Zoom room with his Peloton stationary bike in the background. He was overseeing the COVID-19 Prevention Network, which recruited, vaccinated, and collected data on the clinical trial volunteers. This network accounted for about a third of the sites used in Warp Speed; the rest of the trial volunteers would be seen at commercial sites. Despite being just one of many stakeholders, Corey tended to dominate conversations and was unbending about his own scientific vision. He saw Warp Speed as a grand experiment that had to meet the highest standards. He meant to do right by the American people, who were bankrolling these trials. Everyone agreed that Corey knew what he was talking about and that many of his goals were worthwhile — imperative, even — for the furthering of science. Yet several players inside the Humphrey Building would come to feel that some of his demands went beyond the scope of the ambitious and time-sensitive operation that the president had backed and that the companies had agreed to. The stakes were high; the tensions higher. Finding that middle path was going to be challenging.

In early June, Mascola had briefed Moncef Slaoui on the NIH's plan for "harmonizing" the clinical trials. Moderna's planned phase 3 trial start date, July 9, was approaching, and there was no time to lose. Because of the adverse reaction experienced by Ian Haydon and several others, the company had dropped its highest dose, 250 micrograms, from consideration in its upcoming trial. The Goldilocks dose—not too high, not too low—seemed to be somewhere between 25 and 100 micrograms. In its phase 2 trial, the company was testing a 50-microgram and 100-microgram dose on several hundred individuals.* In order for Moderna to begin phase 3 clinical trials in less than a month, Operation Warp Speed had to sign off on the trial protocol, a hundred-plus-page document that included details ranging from the technique by which volunteers would be randomly assigned to the placebo or vaccine group to how and when the data should be analyzed.

The other four companies that had, at least in principle, joined Operation Warp Speed would have to undertake the same process to get sign-offs on their trial protocols. AstraZeneca was aiming for a July start to its trial; Johnson and Johnson was gearing up for early fall; cash-poor Novavax and venerable Sanofi were a distant fourth and fifth place. Notably absent from Operation Warp Speed's harmonized trials: Pfizer. The unstoppable multinational was likely going to be the second company to enter phase 3 clinical trials in the United States, but, as chief scientific officer Mikael Dolsten had already made clear to Francis Collins, Pfizer would be taking no money for development. The company would be assuming all the risk and, it hoped, reaping all the reward. It would soon offer Operation Warp Speed a chance to lock in its two-dose mRNA vaccine at an exorbitant one hundred dollars per dose—that meant ten billion dollars to vaccinate fifty million people, less than a sixth of the U.S. population. That was more than double the sum Moderna had floated. Moncef Slaoui and General Perna declined.

For Operation Warp Speed's closer allies and their clinical trial plans, Slaoui crafted a two-page memo in all caps of the "KEY TENETS" of the Operation Warp Speed clinical trials, cementing the terms that the companies—or *sponsors*, in FDA lingo—would have to follow in order to fall into the harmonization scheme. He planned to have a call with all five companies at once that week in order to finalize the ground rules

---

* Moderna also included a placebo group in its phase 2 trial, though including placebos in phase 2 trials is not standard.

for participation. "This way there is no misunderstanding," he wrote Mascola. One of the most important of his key tenets — tenet no. 5 — related to what happened when volunteers in the trial tested positive for COVID-19. "ENSURING OPTIMAL CARE FOR SUBJECTS INFECTED IN THE TRIAL WILL BE COMMON AND AGREED BY NIH/NIAID/OWS AND THE SPONSOR."

It was easier said than done. Recall that Moderna had been skipping toward phase 3 clinical trials funded by BARDA when Operation Warp Speed was still coming together. Moderna's chief medical officer, Tal Zaks, had already written a lengthy protocol for the company's planned clinical trial that was now under review by ethics boards and the FDA. For this unproven vaccine maker, which had the threat of Pfizer moving forward with its own mRNA vaccine, the goal was to collect the minimum amount of data needed as rapidly as possible to demonstrate to the FDA that the Moderna vaccine was safe enough and effective enough to merit an emergency use authorization and, down the line, full approval.

During hourlong conference calls, Larry Corey latched onto Moderna's protocol and picked it apart line by line, repeatedly butting heads with Zaks, who was nearly as pigheaded as Corey. John Mascola, who stood at the complex intersection of Operation Warp Speed, Moderna, and the NIH, tended to agree with Corey but would mediate during backchannel negotiations with Moncef Slaoui or Moderna CEO Stéphane Bancel.

By the middle of June, the discussions with Moderna were breaking down. One particular point of contention: Corey believed that volunteers in the study who came down with COVID-19 should receive continuous temperature and blood-oxygen monitoring. He had heard stories about undiagnosed low oxygen levels causing heart attacks in some elderly people. He felt that anyone who tested positive had to be tracked closely as the disease developed. Being able to classify volunteers who developed mild, moderate, or severe COVID-19 would also provide a more granular evaluation of how well the vaccine worked. What if it didn't stop infection but could prevent severe disease? Tal Zaks replied that oxygen monitoring was not part of the standard of care for all people infected with COVID-19, many of whom didn't show serious symptoms. It didn't make sense to impose it on Moderna. Corey believed that clinical trial volunteers deserved better than the standard of care: "They're not just an end point, they're a human being," he said.

During heated debates like this over the coming months, those on the Operation Warp Speed team inside the Humphrey Building would get quiet and let Corey rant, figuring they could come to terms with Moderna offline. The NIH was advising the operation; it wasn't running the show. But sometimes Slaoui would hit the mute button inside the conference room in the Humphrey Building and urge his team on. "Speak up, this guy is terrorizing everyone!" he said one time. Then there was the day Corey and his academic partners popped up on Warp Speed's video screens during a vaccine portfolio meeting — the business-sensitive discussions that were supposed to be walled off from the meandering technical discussions. Slaoui was irate afterward. The Operation Warp Speed folks made sure *that* didn't happen again.

The oxygen-monitoring debate was not getting resolved. "We need to come to agreement on this," Warp Speed's vaccine lead, Matt Hepburn, said one day. Toward the end of June, Slaoui asked his team for an estimate from Moderna about how long the protocol changes Corey was demanding would delay the trial. The word came back that Corey's changes would slow the trial start date three weeks, from July 9 to July 30, as the protocol was reviewed again by the FDA and the institutional ethics boards. Consternation ensued. It was a public health emergency, after all. In the end, it was agreed that volunteers who came down with COVID-19 during the Moderna trial would receive daily, though not continuous, oxygen monitoring. Was it the right call? Larry Corey called it a win.

In the two months since Peter Marks and Bob Kadlec had initially proposed Operation Warp Speed, it had become clear that the coronavirus was not going to fade away over the summer, which was, in a perverse way, good news for the pandemic response.

A successful phase 3 trial had to be run in places where people were likely to be naturally infected by the coronavirus. That meant that the more coronavirus in a community, the more quickly you'd see a difference between the outcomes of those who had received the vaccine and those who had received the placebo. The challenge was twofold: The virus was not evenly distributed around the country, and it was hard to evaluate community spread. Coronavirus test results were still limited, and the CDC's database on hospital usage was clunky and incomplete, making it difficult to track the outbreak with the needed precision.*

What were the best clinical trial sites to use to achieve success as quickly as possible? Julie Ledgerwood, one of the Vaccine Research Center's deputy directors, had begun her career at the NIH working in Barney Graham's lab and had taken over his early role running all the VRC's clinical trials. Now she needed to answer that question — to come up with a targeting strategy to identify the sites around the country where Warp Speed would be recruiting and following clinical trial volunteers.

One of Ledgerwood's first ideas was prisons. Prisons had been ravaged by COVID-19 outbreaks, and some prison doctors encouraged her to run the vaccine trials on their inmates. A bioethics evaluation the NIH performed by the June 8 indicated it could be acceptable to do that, but the practical issues were daunting. Inmates would need smartphones, for instance, to report their symptoms. Were wardens going to agree to that? Besides, the prisons were so heavily hit that many inmates already had antibodies.

Ledgerwood's second idea was Google. She had spoken to the company, which was able to predict imminent outbreaks based on searches for terms such as *COVID symptoms*. The problem was that these forecasts could give you an accurate picture only a few weeks into the future, while the process of setting up a trial and enrolling patients would stretch for months.

It was around this point that Jose Arrieta, the chief information officer at HHS, told Ledgerwood about an initiative he was working on with ASPR and the CDC called HHS Protect that would provide real-time data on testing results and hospital usage. From the beginning of the pandemic, Bob Kadlec had wanted to be able to stand in the Secretary's Operations Center and watch the outbreak like an approaching hurricane. "Tell me where the next outbreak ('brush fire') will occur? . . .

---

* There's a joke among state health officials that the letters *CDC* stand for "Can't Do Computers."

Where are the vulnerables? . . . How can we surge capabilities within 72 hours to target areas of impending risk?" he had written in a May 3 slide for his staff. That investment, Arrieta realized, might pay off for vaccines as well.

He connected Ledgerwood to another guy he was talking to, Jeremy Achin, an insurance actuary turned artificial intelligence guru whose company, Data Robot, was developing a tool to predict COVID-19 hot spots. Ledgerwood thought the company sounded like it made video games, but when she saw what Achin was capable of, she knew it was exactly what the clinical trials team needed. His algorithm would integrate mathematical models of disease spread with information on social-distancing measures. It would then correct itself each day based on the latest data, a process known as machine learning.

Ledgerwood and her small team, the Predictive Analytics Working Group, began manually inputting the GPS coordinates of over a thousand possible clinical trial sites with the coordinates for around twenty thousand facilities around the country where outbreaks were likely to occur: nursing homes, prison systems, even Amazon distribution centers. There were also emerging hot spots in Missouri, Texas, and Utah, states whose governors had been openly hostile to social-distancing measures and mask requirements. One of the things Ledgerwood noticed from Data Robot's output was that, although the outbreak seemed to be moving from the coasts to the center of the United States, the pattern didn't look as if it would hold up in the future. By late fall and winter, it was clear: the pandemic was going to blanket the entire country.

Something else was clear: it wasn't going to affect everyone equally.

Francis Collins, the NIH director, was unsettled by the way the country was being torn apart. Social justice had always featured in Collins's thinking. The May 25 killing of George Floyd, an unarmed Black man, by police in Minneapolis horrified both him and his wife, Diane Baker, a genetics counselor. The video of Floyd pleading for help and calling for his mother before passing out under the knee of Officer Derek Chauvin was too much to bear, even for a nation that had seen this kind of brutality time and time again. It was murder in broad daylight. Across the country — hanging from windows, painted on walls, written on T-shirts — you saw the letters *BLM*. Black Lives Matter. Amid the pent-up suffering of a populace already struggling with lockdowns and economic hardship, Floyd's killing was the match tossed into the powder keg. From Los Angeles to Chicago to Portland, there were protests. These began peacefully but some turned into riots, with cars set on fire and storefronts smashed.

On June 1, Collins sent a message to the "NIH Family" after the killing, telling them that he was "heartbroken by the injustice of this dark moment." The institution needed to address "the health disparities that prevent many from experiencing the full and complete life they hope for and deserve." He ended his message with a quote from Martin Luther King Jr. that seemed to speak to the particular relevance of social justice in the time of COVID-19: "We are caught in an inescapable network of mutuality, tied in a single garment of destiny. Whatever affects one directly, affects all indirectly."

Four days later, he and his wife grabbed their face masks and a couple of protest signs and headed to the pedestrian mall on Sixteenth Street in Washington, DC, which had recently been renamed the Black Lives Matter Plaza. There, they knelt for eight minutes and forty-six seconds of silence, the amount of time (it was then thought) that it had taken for George Floyd to die. Afterward, Collins remained shaken. He knew it was critical to press Operation Warp Speed to double down on its efforts to include the Black community in the vaccine trials. Black people were among those suffering most from COVID-19, and there was a long history of them being underrepresented in clinical trials. If the NIH didn't work to rectify this, Collins understood he would have failed in his duty to the country's citizens.

On the evening of Friday, July 3, Collins began an e-mail with an apology for ruining everyone's Independence Day holiday. He was requesting an urgent Zoom call for the NIH team the next afternoon. "Given how the burden of this disease has fallen so heavily on African Americans and Latinos, as well as older people and those with chronic disease, it will be essential for the volunteers who take part in these trials to represent those demographics effectively," he wrote to Tony Fauci, John

Mascola, Larry Corey, and the other members of the NIH team. "We are also concerned that the companies involved may resist this mandate, since it will potentially slow the trial and cost additional funds."

During the call, the team discussed ways to increase minority recruitment through organizations like the National Black Church Initiative, the Congressional Black Caucus, and the National Hispanic Medical Association. The participants also proposed bringing celebrity voices into the effort. NBA superstar LeBron James, perhaps? Or how about Oprah Winfrey? Lin-Manuel Miranda, the creator of the musical *Hamilton*, made the list, as did singer Jennifer Lopez.

In the days after the call, Mascola and Corey brought up the concern to Moncef Slaoui's team. Slaoui hadn't fully considered the issue but the entire Warp Speed team felt it was worthwhile. On this matter, Moderna's Tal Zaks agreed. In an addendum to Moderna's clinical trial protocol, Zaks added a single sentence: "Given the disproportionate disease burden of COVID-19 in racial and ethnic minorities, the study will also aim to enroll a representative sample of participants from these minority populations and adjust site selection and enrollment accordingly." The wording, notably, was not binding.

19

THREE HUNDRED
MILLION DOSES

General Gus Perna knew that no matter how successful the mission was, those three hundred million vaccine doses weren't going to materialize all at once. Manufacturing would ramp up slowly, in fits and starts, and the first people to receive the shots in the fall or winter would be, at least in some quarters, objects of envy.

For his operational planning, Perna needed a cue from policymakers and public health experts about who was going to be first in line. Would the first shots go to those most at risk for developing severe COVID-19, such as the elderly and people with diabetes or obesity? What about the nation's twenty million health-care workers or the sixty million essential workers harvesting the nation's food or running the public transit systems? A high proportion of the workers keeping society functioning were minorities, who were more likely to develop severe COVID-19. The problem was that if you prioritized everyone, you prioritized no one.

Considering that such decisions would be momentous, NIH director Francis Collins wanted to be part of the process. His first recommendation was to commission a study by the National Academy of Medicine to develop the initial framework for fairly allocating vaccines.

There was some head-scratching down at the CDC in Atlanta as to exactly what role this independent body would take on. Was it going to infringe on the CDC's turf? Or was it just going to serve as a kind of think tank that would help inform the CDC as it began its own discussions and came to a decision? If it was just the latter, the CDC staffers supposed, they could live with it.

Only after the first meeting of the National Academy of Medicine committee, however, did the participants realize they were missing a key expert—a gerontologist! Nursing homes and assisted-living facilities, after all, were being hammered by COVID-19. They accounted for an estimated 40 percent of deaths from the disease. The committee soon brought on Michael Wasserman, the medical director of the Los Angeles Jewish Home, which cared for over a thousand elderly at its facilities in Southern California. Back in February, when Wasserman heard about that first, devastating outbreak at the nursing home in Washington State that ended up killing twenty-two residents, he was inconsolable. "No matter what effort we put forth, this virus is going to find its way into our facility," he told his staff.

Wasserman joined the committee's virtual meetings over the summer and its members were in broad agreement that one of the initial goals of Operation Warp Speed had to be to protect nursing-home residents as well as the frontline health workers who were most likely to come in contact with the sick. As vaccine production ramped up toward the end of 2020 and the beginning of 2021, inoculations would expand to other groups. For a vaccine to achieve its full potential in ending the pandemic, it was thought that the population needed to cross a threshold known as "herd immunity." Vaccinating people was like thinning a forest to prevent wildfires. If enough people were immune to the coronavirus, then an isolated spark would no longer be able to cause a conflagration. The more contagious a virus is, the higher the threshold for herd immunity. To stop the spread of the coronavirus, epidemiologists estimated at the time that around 70 percent of the U.S. population should be vaccinated or have antibodies from a real infection.

As the head of the CDC's COVID-19 response, Nancy Messonnier of the National Center for Immunization and Respiratory Diseases was going to be a key player in ensuring that doses moved from the factories to the front lines. A brilliant scientist who was a bit of a control freak, Messonnier had taken a reputational hit from the testing failures of her organization early in the pandemic, but now she had another chance to have a positive impact in the COVID fight. While Perna's team, overseen by Paul Ostrowski, would figure out how to deliver doses to the states, she would work with state health agencies to get them the last mile. The two systems needed to meet in the middle. Messonnier told the Operation Warp Speed team that she wanted the DoD's help to develop a plan for working out the fine-bore distribution details in two or three states.

On the afternoon of July 9, Deborah Birx and others advising the Warp Speed effort took part in an all-hands-on-deck vaccine-distribution meeting in the Eisenhower Executive Office Building, across the street from the White House. "What will the division of roles be between the federal, state, and commercial distribution channels?" read an e-mail sent out beforehand. "Who do we anticipate incurring the financial costs for vaccines distribution and administration for a variety of potential recipients?"

Messonnier called in from Atlanta to present the CDC's proposed strategy, which was built, in part, on its distribution of the swine flu vaccine in 2009. That rollout had piggybacked on the tools and relationships the CDC had developed through its Vaccines for Children Program, she said. State, local, and territorial health systems could order doses from that system and distribute them to clinics and doctors' offices. The CDC could track every dose administered and to monitor for adverse events.

Jose Arrieta, the chief information officer at HHS, asked how many total vaccination sites there were in the United States. Messonnier responded that there were approximately 250,000 locations, ranging from retail drugstores and urgent care centers to full-fledged hospitals. It sounded good, but then she admitted that the CDC had connectivity to only about 45,000 of those through its childhood vaccination program.

"How are you going to close the gap?" Arrieta asked.

This was hitting on a touchy issue. HHS had already scoffed at Messonnier's request for five to six billion dollars to help states' cash-strapped health departments manage distribution and administration. Local health departments and hospitals had already been allocated $175 billion for COVID-19. Why couldn't that be used in the vaccination drive? But the bottom line was that Warp Speed and its backers inside HHS headquarters foresaw a smaller role for the states and a larger role for the private sector.

Birx wasn't pleased with the Warp Speed strategy laid out in Messonnier's presentation. The CDC, she said, needed to make sure it got the vaccine to community health centers and mass vaccination sites at churches and schools, which was how the country had quashed the polio epidemic in the 1950s. "If you push everything to the hospital system, you're going to miss the most important part of the population," she warned the team.

Those were all ideas that the CDC was already exploring, at least internally. "We're presenting this because this is what Warp Speed told us they wanted," Messonnier said finally. Birx shot a nasty look at Ostrowski, the top Warp Speed representative at the meeting.

The meeting ended with Birx saying that she wanted to see the distribution and administration plans in greater depth and that she wasn't going to give the teams the green light until that happened. Messonnier and her team were surprised to hear that Birx was claiming dominion over the CDC's traditional role, and they felt more confused than ever about the lines of authority with Operation Warp Speed and what their next steps should actually be.

For Alex Azar, good news or, at a minimum, no news meant he had another day on the job, and the best news of all was an appearance on *Fox and Friends,* one of Trump's favorite programs. On July 22, Azar offered a juicy scoop to Fox hosts Steve Doocy and Ainsley Earhardt.

"Breaking news, right here on your show," Azar said, "which is that under the president's leadership we just signed a contract with global pharmaceutical leader Pfizer to produce a hundred million doses of vaccine starting in December of this year." After a month of haggling, the government's final terms with Pfizer came to about twenty dollars per dose, or $1.95 billion dollars for a hundred million doses.

CEO Albert Bourla tried to get General Perna and Moncef Slaoui to commit to more, but they shrugged him off. Not knowing which vaccines were going to succeed, Perna and Slaoui had to balance out the risk among the five other vaccine manufacturers apart from Pfizer. One of the major problems on the Pfizer end, from Perna's perspective, was that the company was unwilling to share with the government many details about its production capacity and timeline beyond those first doses. Perna warned Pfizer that without more information, he would not be able to help them out by prioritizing orders under the Defense Production Act. He couldn't put the operation at risk of Pfizer using the government to siphon off resources from other competitors.

After his appearance on Fox, Azar didn't get much out of Trump. Later in the day, he saw him in person inside the White House Cabinet Room during an unrelated meeting about drug pricing. Trump's chief of staff, Mark Meadows, and Jared Kushner were among those seated around the table. Trump decided not to go with one of Azar's beloved proposals. As often happened with the president, the ideas of the day would accumulate in his head and rattle around like nickels, eventually emerging all at once like the payout from a slot machine. "We're giving all this money to drug companies," Trump said. "Pfizer! They're getting two billion. Who gets to approve this?" he asked, turning to Meadows. "Mark? Do you get to approve this?"

"Mr. President," Azar cut in, "I approved it."

Trump turned to his health secretary and wrinkled his brow. "Wow, I wish I had *that* job." He looked around the table at the others. "Look at the power this guy has! Two billion dollars to Pfizer, and we may not get anything."

Meadows defended Azar. "We always knew we would spend money we wouldn't get anything for," he said.

Azar explained that the Pfizer deal didn't obligate the government to pay a cent if the company didn't deliver one hundred million doses of FDA-approved vaccine by December. "Mr. President, what *exactly* is your problem with that?" he growled.

Kushner explained what a coup it was to negotiate down the price. "It's an amazing deal." Trump became quiet after that.

Two days later, Pfizer chief scientific officer Mikael Dolsten listened intently as the vaccine team explained its dilemma to him and CEO Albert Bourla. Dolsten was in his office at his summer home in Westhampton on Long Island, the backyard pool calling him on this hot July afternoon.

In order to stay ahead of Moderna and have a chance to complete Pfizer's trial before coronavirus numbers worsened in the winter, the company had to make a final decision about which of its vaccine candidates to take into phase 3 clinical trials. "We need to make a decision today," Dolsten told his colleagues at both Pfizer and BioNTech in Germany.

The company had initially started testing four vaccine candidates in humans back in May. Dolsten and the rest of the team had been enthusiastic about their self-amplifying RNA vaccine candidate. It not only stimulated the cells to produce the spike protein, as Moderna's vaccine did, it also stimulated them to produce copies of the RNA itself. In theory, this would require much smaller doses of vaccine, easing the potential manufacturing crunch with a national or global scale-up. But if RNA vaccines were already a technological novelty, self-amplifying ones were even further in the future; they involved a more complicated sequence and more unknowns. So Dolsten's team axed that one. A second candidate, which also used the complete spike protein, had not made use of Katalin Karikó's modified mRNA strategy, and it generated too strong a reaction in people during the phase 1 clinical trial.

This left the final two: BNT162b2, the full-length stabilized spike protein nearly identical to the Moderna vaccine sequence, and BNT162b1, which was the business end of the spike, known as the receptor-binding domain. Kathrin Jansen, Pfizer's head of vaccines, explained that the full-length spike had the potential for greater efficacy. The company, however, had more data on the binding-domain vaccine because it had entered phase 1 clinical trials a couple of weeks earlier. In particular, the results from older adults were looking very promising. Unfortunately, the team didn't have the comparable data from the full-length spike yet. To save the most lives, everyone knew, their vaccine absolutely had to stop the virus in older adults, whose aging immune systems were not as good at storing a "memory" from a vaccine or a previous infection.

One contingent of the team strongly favored going with the receptor-binding-domain vaccine. In addition to having a fuller picture of the data for that version, the team members argued that the fragment was more unique. Every other company, including Moderna, was going with the full-length spike. "Why would everyone be picking the larger fragment?" Dolsten wondered. Both Dolsten and CEO Bourla had attended a training session with Daniel Kahneman, the Nobel Prize–winning psychologist, who had helped them understand their cognitive biases. It was natural that the team was attached to its first candidate, but there was no real reason why Pfizer needed to be different. The competition was really all about speed and scale.

Dolsten and Bourla gravitated toward BNT162b2, the full-length stabilized spike, and they swayed the rest of the team, albeit with considerable uncertainty in everyone's mind. That evening, July 24, Pfizer submitted its application to the FDA to proceed with its phase 3 clinical trials — just days behind Moderna.

A little after six a.m. on July 27, Dawn Baker pulled into the parking lot of Dr. Paul Bradley's office in Savannah, Georgia. Baker was a local news anchor on WTOC-11 who had seen the signs around town imploring residents to "Join a COVID-19 Vaccine Study." The advertisements, paid for by Meridian Clinical Research, promised participants up to $1,500 for volunteering. Baker, a Black woman, had even introduced a segment about it on the evening news.

On this steamy Southern morning, she passed under oak trees festooned with Spanish moss and walked through the office complex, consisting of a series of stately two-story red-brick buildings, until she found the door to Bradley's office. Baker took a seat in an exam room and bared her left arm. Moderna's vaccines were stashed inside a stainless-steel freezer with a sign that read CRUSHING COVID. "You ready?" asked a woman in a white lab coat.

Baker said without hesitation, "Yes, ma'am." The woman rubbed her upper arm with an alcohol-prep pad, then injected Baker with her first dose of the Moderna vaccine, making her the first volunteer in a phase 3 coronavirus vaccine trial. When her own television station asked Baker why she'd volunteered, she replied, "I think it's really bigger than me, but what I hope people see is that we all are important and we all can be a part of solving this."

The responses from the Black community in Georgia rolled in on Facebook. Some commenters were positive, praising Baker for her bravery and wishing her the best. Others shared dark memes, such as a video of a group of six men at an African funeral procession dancing with a coffin on their shoulders.

"Oh girl, no you didn't," read one comment.

"She going to die," read another.

"I would never ever risk my life with something coming from Donald Trump."

"So if it was Bill Gates foundation vaccine you would have been first in line? Ok. Got it."

"No vaccines for my family."

"I can't help but feel like this is another Tuskegee experiment."

The Tuskegee syphilis experiment came up a few times in the Facebook comments, a reference to the notoriously unethical study run by the U.S. Public Health Service and the CDC from 1932 to 1970. The idea behind the study was to understand the effects of untreated syphilis, which infects the nervous system and other tissues, causing blindness, mental illness, and death. In order to recruit a group of six hundred sharecroppers to participate, researchers lied to them about the purpose of the study and promised them free medical care, even though many were never provided with a diagnosis or treatment, putting not only them but also their sexual partners at risk. When the existence of the study leaked out to the wider medical community, doctors expressed outrage and the story became front-page news in the *New York Times* on July 26, 1972. The mistrust that the Black community felt for white doctors seemed irreparable, and, fifty years later, many were still reluctant to participate in medical research of any kind.

Hours after Baker became Moderna's first volunteer, Pfizer vaccinated its first. In one of those strange coincidences in life, Barney Graham's old roommate from Rice, Bill Gruber, was overseeing Pfizer's phase 3 clinical trial. The two men had both gone on to medical school and overlapped at Vanderbilt in the 1990s before Graham went to the NIH and Gruber entered the corporate world. Gruber was launching a clinical trial like none other, with one hundred and thirty sites around the world that had been established earlier for the testing of Pfizer's blockbuster pneumococcal vaccine. The company had wanted to use some of the sites at certain academic institutions that were also part of the

COVID-19 Prevention Network, but Larry Corey and John Mascola shot them down — only one mRNA vaccine at a time, they said. Pfizer complained to Mascola about it, and he told the company it was welcome to use the sites after Moderna's last volunteers got their second doses.

Thanks, but no, thanks, Pfizer replied. Every decision that went into Pfizer's protocol seemed designed to deliver the company a speed advantage over Moderna. The Pfizer timeline involved giving participants the second shot after three weeks, compared to the four weeks used by Moderna. And while Moderna wouldn't start counting virus cases among its cohorts until two weeks after the second dose, Pfizer would be doing it at just seven days afterward. Finally, Pfizer would analyze its interim results when its trial had accrued thirty-two cases of COVID-19. Under Moderna's protocol, that trigger was fifty-three cases. The FDA's statisticians strongly advised Pfizer to be more conservative, but Pfizer had a race to win.

Before the historic day came to an end, there was a hitch. A group of advocates from the HIV community posted an open letter to NIH director Francis Collins expressing anger about being excluded from the trial. This was one issue on which Larry Corey had failed to prevail during the Operation Warp Speed meetings. There are always exclusion criteria in clinical trials, and avoiding testing a new vaccine on people with compromised immune systems made a lot of sense. That was especially true amid a pandemic when you wanted to reach the finish line — deploying a vaccine for the widest swath of the adult population — as quickly as possible. An adverse event in one of those individuals could cause the clinical trial to come to a screeching halt. One could always go back and run subsequent studies with special populations, as was usual with pregnant women.

Corey had clued these advocates in to what was happening. His clinical trial network had had a relationship with them from years of HIV studies, and the activists decided to mobilize the moment that details of the coronavirus vaccine study became publicly available on the clinical trials website. "The HIV community was promised that there would be community representation in all aspects of COVID-19 clinical trials development," they wrote. "But all we see is procrastination which now appears to be purposeful." They demanded that Collins do everything in his power "to intercede with Moderna to reverse this exclusion and ensure that such an exclusion never happens again." There was a collective sigh inside the mission-focused team in the Humphrey Building. "They're trying to interfere with an ongoing trial," one team member complained. Within a week, Moderna put out that fire, releasing a statement that the company would change its protocol and enroll HIV-positive volunteers after all. Corey had emerged victorious once again.

Over the summer, the rift had grown between General Perna's team in the Humphrey Building and the CDC in Atlanta. It was one part politics and one part culture clash.

For months, the CDC's top leadership had felt steamrolled by various factions of the administration. Not long after that July meeting about distribution, Deborah Birx directed hospitals to start reporting their data directly to the newfangled HHS Protect system rather than to CDC's own legacy database. CDC staffers contributing to the project in Washington summarily checked out of their hotels and flew back to Atlanta. When it came to the vaccine-administration plans, CDC officials were kept from sharing all manner of information directly with the states. They were even denied the ability to set up a website. The end result was that the CDC staff started covertly passing vaccine-related slide decks to state health officers through preexisting folders they had established on an application called ShareDrive.

There were chilly phone calls here and there in the coming months, but a true collaboration at an agency level? Not really. Instead, it would come down to the individuals working on the effort, people who were now thrust together under the same roof, driven by a shared sense of mission and a desire to keep moving forward no matter what. Anita Patel, a CDC pharmacist dividing her time between DC and Atlanta, was instrumental in setting up a partnership with retail drugstore chains, such as Walgreens and CVS, that included a program specifically for administering vaccines at nursing homes. That federal-level program would complement the CDC's plans to get vaccines out to clinics and hospitals through the states.

Patel and her group, however, were caught off guard by the intensity of the questioning from Warp Speed's Paul Ostrowski. Ostrowski had a buzz cut and the demeanor of a drill sergeant. He would fly off the handle if someone didn't have a number ready when he asked for it or if his computer started making annoying noises. During planning meetings, he would ask rapid-fire questions: *How much dry ice can you actually put on a plane? Where will the shipping labels be printed? What time does that clinic open?* The public health experts sometimes felt like they were under attack and wondered if they were supposed to listen to the military or if the military was supposed to listen to them. Trust was at an all-time low, and they had no one back in Atlanta to help them navigate this minefield.

It wasn't any easier for the military folks, who were trying their best to learn about this new and unfamiliar world of public health. Some expressed growing doubts about achieving the goals for the rollout. At six p.m. on July 22, General Perna convened a virtual huddle with just the military personnel under his direct control to deliver what is known on the battlefield as the commander's intent. "Our mission is clear: three hundred million doses by the end of the year," he said. Every member of his team had been handpicked for a specific purpose. They would have to develop options and identify the greatest risks of the operation. "I'm tired of seeing the team saying we will not get it done," he said sternly. "Gotta figure it out, get it done."

He said that because his team was "not in charge," they would have to lead from behind. He went on: "Our approach has to be to figure this out. And it must be agile, adaptive, and innovative." He repeated those final words — *agile, adaptive, innovative.* "No more crap about what we're going to accept! No more pacified response! The only thing I'm interested in is the end state!" He finished his bit as he would on a battlefield radio: "Over."

20

PLASMA DRIVE

On the morning of Saturday, August 22, Secretary Alex Azar got a phone call from the president. "Did you see my tweet?" Trump barked. "Why hasn't the FDA approved convalescent plasma? Is it that fucking Hahn again?"

Hahn was Stephen Hahn, the FDA commissioner. A bald man with a white goatee, Hahn was a Washington outsider, a likable radiation oncologist who had guided M. D. Anderson Hospital in Houston out of a financial crisis. Following his appointment as the commissioner in December 2019, he had developed an unusually cozy relationship with the White House, in part due to the pandemic. Not only did he make regular appearances on the coronavirus task force, he provided weekly briefings to John Fleming, a doctor in the White House who advised Mark Meadows, Trump's chief of staff. A person's standing with Team Trump, however, could shift at any time.

Azar didn't regularly read the president's tweets, though he usually heard about the more inflammatory ones pretty quickly. He scanned through Twitter on his iPad and found what the president had written: "The deep state, or whoever, over at the FDA is making it very difficult for drug companies to get people in order to test the vaccines and therapeutics. Obviously, they are hoping to delay the answer until after November 3rd. Must focus on speed, and saving lives!"

Vaccines were still months away and Trump had been hoping to declare an early victory of sorts in time for the Republican National Convention on August 24, in two days. Convalescent plasma, blood plasma obtained from recovered COVID-19 patients, had the potential to be a crude therapeutic. Dating back more than a century, the strategy was a low-tech way to transfer the acquired immunity from one person to another and might just improve a hospitalized patient's chance of recovering from COVID-19. Operation Warp Speed was overseeing some of the logistics of deploying massive quantities of the stuff under a compassionate-use protocol, but Trump wanted the FDA to go further and issue an emergency use authorization. The tweet was the last thing Azar needed if he wanted to keep the peace inside his organization.

"For the love of God, stop doing this!" Azar said. "We talked about this." He reminded Trump that it was going to make it look to the public as if the FDA bent to the president's will. "The FDA are bureaucrats and are not going to react well," he added.

Azar knew he had to pick his battles to prevent a revolt among his agencies. Days earlier, HHS had effectively stripped the FDA of its authority to regulate certain coronavirus tests conducted by private labs, which had already put Hahn in the hot seat with his career staff and made Azar look like a bully. Azar told Trump not to worry, that the FDA was, in fact, working on an EUA. It might well be signed tomorrow morning.

"Oh, good, good," Trump said. "I want to announce it. I'm going to call Kayleigh. We're going to set up a press conference." Kayleigh McEnany, the president's glamorous press secretary, was in Tampa, Florida, and quickly hopped on a plane back to DC.

Never mind that the FDA still didn't have a final answer on convalescent plasma. Peter Marks, the head of the Center for Biologics Evaluation, had been in favor of an emergency authorization, but when NIH director Francis Collins was consulted on the matter, he balked at the quality of the data being used to support it, which resulted in a public spat between the two agencies. The White House then placed its thumb on the scale. There were the calls from Jared Kushner's A-team to Collins and Marks. "We've really got to get this going," Adam Boehler told Marks that week. More data finally rolled in at seven p.m. on Saturday, and Marks was pleasantly surprised to learn that convalescent plasma showed a statistically significant benefit — at least in a small subgroup of patients. "From my perspective, it is a definite go," Marks wrote Hahn in an e-mail that evening, sharing the analysis.

The ceremonies touting the EUA began the following afternoon. Marks joined Collins in the Roosevelt Room, a conference room in the West Wing with a long wooden table and a portrait above the fireplace mantel of Theodore Roosevelt on horseback heading off to the Spanish-American War. Alex Azar, Bob Kadlec, and Michael Caputo, the top HHS spokesperson, were already there. Hahn had just flown in from his home in Colorado.

Eight months into the pandemic, the standard of care for COVID-19 patients included the Gilead drug remdesivir, which had to be used early in treatment, and the steroid dexamethasone for more serious cases. The combination of drugs along with better care, such as the avoidance of ventilators, had reduced deaths among hospitalized patients from 17 percent to 9 percent, but thousands were still dying every day. Convalescent plasma was a small advance, at best, something to be rolled out with a careful and directed communications strategy to doctors and patients.

Azar and Hahn headed across the hall to the Oval Office for the press conference. Marks, Collins, and Kadlec were left in the Roosevelt Room with Caputo. The portrait of Roosevelt, the brave and principled Republican leader, gazed down at them from above the fireplace. Collins wondered what the holdup was.

"Oh, they've started," Caputo said, looking at his cell phone. Collins and Marks gathered around Caputo to watch the press conference on the tiny screen. They saw Trump step up to the podium in a celebratory mood and begin by saying he was looking forward to the Republican National Convention. He said he was announcing something he had "been looking to do for a long time" and then touted convalescent plasma as "a very historic breakthrough in the fight against the China virus."

Next, Azar, in a sky-blue tie, gave a nod to the FDA's Marks and Hahn for their hard work. "We saw about a thirty-five percent better survival in the patients who benefited most from the treatment, which were patients under eighty who were not on artificial respiration," he said. "I just want to emphasize this point because I don't want you to gloss over this number. We dream in drug development of something like a thirty-five percent mortality reduction. This is a major advance in the treatment of patients. This is a major advance." He went on to promise more drugs coming out of the pipeline thanks to "the president's Operation Warp Speed," which would take the new drugs all the way through manufacturing, "so that if they meet FDA's gold standard for safety and efficacy, they can begin reaching patients without a day wasted."

Azar's pitch for convalescent plasma was a bit overblown, Marks thought, but that was the game Azar played and it was par for the course if you wanted to survive in the Trump era. The person whose credibility mattered most of all was Marks's boss Stephen Hahn, the face of the FDA, a doctor. He had taken the Hippocratic oath upon his graduation from medical school, promising always to do right by his patients.

Hahn stepped up to the podium. "From the beginning of this pandemic, the president has asked FDA to cut back red tape to try to speed medical products into the hands of providers, patients, and American consumers," he said. He went on to describe for the layperson exactly what convalescent plasma was and explained that his expert scientists had reviewed the "totality of data" and had concluded there was "a thirty-five percent improvement in survival," which he called a substantial clinical benefit. Hahn said that if you had a sample of a hundred people who were sick with COVID-19, "thirty-five would have been saved because of the administration of plasma."

"Oh, my!" Marks exclaimed inside the Roosevelt Room. "He botched it."

Marks and Collins exchanged glances of horror. The 35 percent improvement number that Marks had shared with Hahn and Azar was an estimate of the relative-risk reduction for a subset of patients, not the absolute risk. In reality, convalescent plasma wasn't going to save the lives of thirty-five people out of a hundred. It was going to reduce the relative risk of death in this subgroup by 35 percent, from, say, ten out of every hundred patients to around seven out of a hundred. *No self-respecting cancer doctor would ever talk in terms of absolute-risk reductions!* Marks thought. That would be a miracle. Despite all the care Marks had taken to preserve the reputation of the FDA in a time of crisis, Hahn had blown it. And that meant Marks had blown it.

Azar was oblivious to the blunder. He came back to find Marks and Collins still sitting in the Roosevelt Room. "Peter," Azar said, "come with me." He had asked Jared Kushner if he could bring Marks into the Oval Office, so Marks, for the first time, stepped through the threshold and saw President Trump in close quarters. He was in a jovial mood. Azar told Trump, "Peter is the architect of Operation Warp Speed and the leader of the team that approved convalescent plasma." The president expressed his gratitude to Marks, and Marks, who was no fan of Trump's politics, had to admit he was charmed by his emotional warmth.

Over the next twenty-four hours, however, Hahn's blunder would be picked apart in the press and on Twitter, putting everyone in an uncomfortable situation. The following afternoon at the Republican convention in Charlotte, North Carolina, the president mentioned convalescent plasma, and Hahn knew he needed to send a message to Marks and other members of his staff that he was on their side. At 9:36 that evening, Hahn tweeted out a mea culpa, reiterating the FDA's support of the EUA but also noting the criticism of his comments was "entirely justified."

A week later, though, the furor still refused to die. On August 31, Eric Topol, a cardiologist at the Scripps Research Institute in La Jolla, California, posted a scathing open letter about the convalescent plasma debacle on Medscape, a news service for doctors where he served as editor in chief. "You have one last chance, Dr. Hahn, for saving any credibility and preserving trust in the FDA at this critical juncture amidst the pandemic," he wrote. "Tell us that you are capable and worthy of this pivotal leadership position and that you will not, under any condition, authorize a SARS-CoV-2 vaccine approval before the full phase 3 completion and read-out of a program."

To say that Topol singlehandedly instigated some kind of shift inside the FDA would be giving him too much credit, but his harsh criticism certainly had an impact. Over the previous three months, Peter Marks had gone from the guy who came up with Operation Warp Speed to push vaccines forward at a relentless pace to the guy who had to decide if that pace was *too* relentless. He felt the gravity of his work overseeing the twelve hundred or so dedicated civil servants working at the Center for Biologics Evaluation and Research — an institution that predated the founding of the FDA itself.

CBER had originated as the one-room Hygienic Laboratory, established in 1887, in the attic of the Marine Hospital on Staten Island, New York. At the time, the laboratory, part of the U.S. Public Health Service, had a relatively narrow mandate: testing sailors in order to prevent cholera, yellow fever, smallpox, and plague from being imported into the country. After four years, the laboratory relocated to downtown Washington, DC, and then to a campus in Bethesda, Maryland, where it became, more or less, the National Institutes of Health. Yes, *that* NIH. Until 1972, NIH was the agency regulating vaccines.

The darkest moment for the nation's biologics control laboratory came during the 1955 rollout of Jonas Salk's miraculous polio vaccine. Poliovirus outbreaks had been growing in frequency and size during the 1940s. While most of those infected suffered only mild flu-like symptoms, the virus could also injure the brain and spinal cord. Victims occasionally ended up with a paralyzed leg or arm. In patients with the most severe cases of polio, the muscles involved in breathing and swallowing were paralyzed, and patients either died or spent time in an iron lung — a steel cylinder, six feet long and three feet in diameter, connected to a bellows. In the 1950s, when an average of fifteen thousand Americans were paralyzed each year, parents were afraid to let their children outside in the summer. Ventilators were in short supply. Officials restricted travel and imposed quarantines in places where polio cases occurred.

Salk's vaccine couldn't come soon enough. In an early trial, he found that it was possible to protect against all three strains of polio using a vaccine made from virus that had been killed with formaldehyde. Once those results were released to the press, there was no turning back, and the March of Dimes Foundation contracted five companies to manufacture it. On April 12, 1955, the first doses of vaccine were shipped out. Not long after, James Shannon, then an associate director at the NIH, received a call from an official at the health department in Los Angeles. The official had just learned of two reports of polio in children who had received the new vaccine. "He wanted to know what should be done about it," Shannon later recalled.

The mood of celebration turned to tragedy. Eighty children who had received the vaccine somehow contracted polio and spread it to another one hundred and twenty friends and relatives. Fifty-six of the children who had received the dangerous vaccine doses were paralyzed, and five died. These doses were traced to a single company in Berkeley, California: Cutter Laboratories. Salk's inactivation method, which had worked on small batches of vaccine, was not as reliable with larger vaccine lots, meaning that some of the lots contained live poliovirus. For the national biologics lab, the worst news came later: Prior to the vaccine's release, a government virologist named Bernice Eddy had tested the Cutter vaccine on eighteen monkeys and found that some of them showed signs of polio and developed paralysis. The head of her laboratory, however, failed to share her findings with the vaccine licensing committee.

After vaccines became part of the FDA's purview, in 1972, the agency would suffer its own crisis with the government's rollout of the 1976 swine flu vaccine. That vaccine was later linked to Guillain-Barré syndrome, an occasionally serious autoimmune disorder. The risk was minimal but not nonexistent. The CDC has estimated that the chance of developing Guillain-Barré from the swine flu vaccine was one in a hundred thousand. Most clinical trials could never detect a side effect so rare, and fears of a repeat naturally contributed to Peter Marks's Center for Biologics Evaluation and Research taking a cautious stance with new vaccine applications. His scientists recognized that a misstep with a vaccine approval could have a domino effect, harming the uptake of long-standing, lifesaving vaccines already on the market. CBER soon became known for moving at a crawl, making absolutely sure that nothing could possibly go wrong.

During the early days of the pandemic, there had been some fear expressed in the scientific community about the possibility of a coronavirus vaccine leading to enhanced disease — the frightening phenomenon that Barney Graham had studied in the inactivated-RSV vaccine all those years ago. In studies of an inactivated-virus SARS vaccine, scientists had seen signs that enhanced disease could occur, particularly in elderly mice, which had less effective immune systems. "Is this something we need to be concerned about?" Marion Gruber, the head of CBER's vaccine office, asked Graham as early as February.

Graham believed that most of the vaccines being developed in the United States were not a major risk. Five out of the six in Operation Warp Speed had been designed with the stabilized spike structure, which generated a larger proportion of neutralizing antibodies and was thus less likely to produce a dangerous allergic response he had seen in RSV. He suggested some tests that vaccine makers could run to ensure that was the case. They should also submit data evaluating what happened when an animal was vaccinated with a dose so low that it didn't protect it against the coronavirus. Did the animal develop more severe disease? Graham and Kizzmekia Corbett had found in their mouse studies that Moderna's vaccine had cleared this high bar. Those mice that received just a smidgen of vaccine showed no worse symptoms than unvaccinated mice.

In late June, Marks's office released a nineteen-page guidance document for COVID-19 vaccine makers. The agency took up Graham's recommendations, laying out the expectation that those studies be completed prior to large-scale clinical trials. CBER also described what was necessary for a vaccine to be fully licensed following the phase 3 clinical trial. The vaccine, the agency said, should be at least 50 percent effective at preventing COVID-19, meaning that, on average, there should be half the number of new coronavirus infections in people who received the vaccine compared with people who didn't. It also wanted to see at least six months of safety monitoring for adverse reactions, perhaps more in some cases.

Only on the last page of the document did CBER mention the prospect of an accelerated path: an emergency use authorization. But exactly what that meant would be decided on a "case by case basis," considering "the totality of the available scientific evidence." The FDA required proof of safety and efficacy from a clinical trial, but the requirements for an EUA remained undefined, which meant the authorization could be manipulated for political ends. In other words, it was trust — more than science, more than the rule of law — that was needed to keep this entire enterprise functioning. On August 23, some of that trust went out the window.

• • •

To the career scientists inside FDA's vaccine team, the convalescent-plasma situation felt like a preview of what was to come. A vaccine had never been released under an emergency use authorization in this country, and a vaccine was a far more delicate matter than convalescent plasma. It seemed clear that the initial guidance the FDA had provided to vaccine makers was not going to be adequate. As Peter Marks and Marion Gruber knew well, the smallest safety risk, if not caught during a clinical trial, would become a very big deal when hundreds of millions of people were vaccinated.

Gruber's office had been fielding detailed queries from all the vaccine makers about the safety data they would be expected to provide for an EUA when they came out on the other side of a phase 3 trial. The six months of follow-up for side effects that the agency had outlined in June was standard for normal vaccine approval, but it seemed overly conservative during a pandemic. Allergic reactions to vaccines occur within hours of a vaccination, and other serious side effects typically emerge within forty-two days. There was a wide gulf between forty-two days and six months. Where should the FDA put the goal line? "We tried to separate ourselves from whatever politics was going on," Gruber said.

Although there were occasional discussions between the members of Operation Warp Speed and the regulators at the Food and Drug Administration, the two groups kept each other at arm's length. Warp Speed didn't want to appear to be trying to influence the FDA decision-makers, while the FDA didn't necessarily go out of its way to be anything other than a black box.

With the finish line approaching, it was becoming clear to the entire Operation Warp Speed team that they could no longer be left in the dark. A landmark virtual meeting between the two organizations took place at 4:30 p.m. on Tuesday, September 1. Moncef Slaoui and General Perna were seated in a conference room on the fifth floor of the Humphrey Building with a half a dozen members of the Operation Warp Speed team. On the other side of the table, so to speak, was the FDA team, including Peter Marks and Marion Gruber.

If General Perna wanted to be ready for what he called the *N-hour sequence*, a military term for the moment the troops get the notification to deploy, his team and the companies working with the operation needed to understand exactly what the FDA required for a vaccine EUA. The FDA hadn't updated the guidance it had published back in June, but Warp Speed had heard that Gruber's office was floating that forty-two-day number. Was this going to be put into some kind of public guidance?

No, Marks said. The FDA would be sending out letters one by one to vaccine makers and would hold a public advisory committee to discuss the topic.

"Have all six been contacted?" Perna asked.

Gruber said that the FDA was in discussions with all the companies. "The letters will be going out shortly," Marks added.

"Can we see the letters?" Perna asked.

"We can see if one of the sponsors is willing to share," Marks said. The communications with companies were considered confidential, even if those companies were working with Operation Warp Speed.

The conversation looped around to other topics, but Perna was intently focused on the timeline, the delivery, the mission. Before the call ended, he made a final plea. "I want to see the totality of the requirement," he said. The FDA's regulators were unwilling to make any promises.

21

TRIALS

B y the end of the first week of August, the Moderna trial had recruited more than four thousand volunteers. This was when Operation Warp Speed got its first glimpse at the racial makeup of the clinical trial cohort: 79 percent of the participants were white, 12 percent were Hispanic, and 5 percent were Black. In proportion to their numbers in the general population, Black people were represented in the trial at less than half the rate of whites. Hispanic enrollment was lagging too. Those numbers did not bode well for achieving the diverse pool of volunteers the company and the government had promised.

Knowing some of the limitations of for-profit clinical trial sites used by companies, the NIH team had lobbied hard for the vaccine trials to be run partly at the academic HIV clinical trial sites that were now part of Larry Corey's COVID-19 Prevention Network. During HIV trials, those academic sites had proved crucial in ensuring that low-income communities and people of color were included. Speed was not their strong suit, however. Each site was part of a separate hospital or university, and agreements and ethics board approvals would have to be obtained piecemeal. Many of the sites had limited space, so General Perna tasked his former post at the U.S. Army Materiel Command in Redstone Arsenal to deliver trailers to each one to expand their waiting rooms. Bob Kadlec worked with testing czar Brett Giroir to purchase several million dollars' worth of DNA-sequencing machines for the sites so they could diagnose COVID-19 cases among trial participants.

The NIH's John Mascola had assured everyone that using the academic network sites wouldn't slow things down, but in fact only a small fraction of them were ready to go by the time the trials were started. On Monday, August 24, Moderna's for-profit sites had enrolled 12,741 volunteers compared to just 976 from the COVID-19 Prevention Network. Not surprisingly, the percentage of Black participants had barely budged. For Moderna to complete enrollment with demographics representative of the American population, Mascola knew that 13 percent of the participants needed to be Black. That meant the company had to recruit 3,250 Black volunteers, nearly ten times as many Black volunteers as it had enrolled in the past month.

Mascola and Warp Speed's Matt Hepburn held a call with Moderna president Stephen Hoge and his team to discuss the issue. Mascola told Hoge that Moderna had to increase the percentage of Black volunteers at least to the double digits. During the meandering hourlong conversation, Hoge was not receptive to making any changes to the trial. No surprise after all the drama with Corey over the clinical trial protocol. Moderna had been vaccinating nearly a thousand people per day, which meant it could hit the thirty-thousand-person mark of full enrollment in two weeks. In order to recruit from more Black and brown communities, the company would have to slow recruitment of whites.

Mascola's reply was that Moderna was shooting itself in the foot. The success of a vaccine trial didn't depend solely on how quickly you enrolled your trial but also on who you were enrolling and on how many "events," or COVID-19 cases, you ended up stopping. Black people were catching COVID-19 at two or three times the rate of whites. The predictive analytics group that Julie Ledgerwood led had shown Mascola that enrolling more volunteers from Black and brown communities could give Moderna a final answer three to six weeks earlier. But enrolling more minorities remained daunting to the company. Mascola's call ended with a stalemate. Hoge told Mascola that the specific diversity metrics were not in the contract with BARDA and that it was up to the COVID-19 Prevention Network to move faster.

Dispirited, late that night Mascola sent an e-mail explaining the situation to Moncef Slaoui. Could he could put pressure on Moderna, both as Operation Warp Speed's leader and as a former member of Moderna's board? "I can see this from the company point of view," Mascola wrote. "They want to finish the study and get an answer. But this is likely not in the best interest of the study or the product, or overall public acceptance of the product." From his experience in the HIV world, Mascola believed that with "focused effort and specific guidance," Moderna could achieve its goals without slowing the trial. "It seems though that Stéphane [Bancel] and Stephen [Hoge] would have to be convinced of this."

Slaoui was fully supportive of bringing down the hammer, but he too failed to make headway with the company. The Moderna leaders were standing their ground on the letter of their contract. Bancel had gone so far as to reach out personally to opinion leaders and pastors in the Black community but otherwise continued to express a sense of no-can-do regarding increasing enrollment. Barney Graham, though he had no official say in these late-stage clinical trials, did what he could to nudge Bancel to try harder during a heartfelt phone call. "I was very forceful with him," Graham said. Bancel couldn't make the decision unilaterally, however. He told Graham that his clinical trial experts needed to be fully supportive.

Mascola reached out to NIH director Francis Collins and Tony Fauci, who would have to answer to the public—and Congress—if the study was biased. "I have done everything I can do. Moderna has basically said thanks, but we are doing what we want," Mascola wrote on August 27. "It's too bad, because the product suffers from too many subjects who are Caucasian and at low risk for COVID-19." Fauci wrote Mascola, "I am always available [to pitch] in with Francis to try and move the needle." It was in fact Collins who moved the needle. He wasn't acquainted with the Moderna team, but this was the issue he felt most passionate about, and when the mild-mannered Collins was passionate about something he didn't hold back. During an August 29 Zoom call with the Moderna leaders, Collins said, "This is unacceptable." If Moderna didn't change course, the company's study would have "no credibility." People were going to accuse them of taking shortcuts. "We are going to offer you assistance, but this cannot be sustained the way you've started."

There was a long silence on the other end.

Over the weekend, Stéphane Bancel and the team at Moderna agreed to make a change. They would pause enrollment at the commercial sites so that the COVID-19 Prevention Network could catch up. It was a win-win — the best strategy for the country and for the company. Investors evidently didn't feel the same way. When word emerged that the company's enrollment timeline would lengthen by a couple of weeks, Moderna's stock dropped by over 3 percent. Pfizer was nipping at its heels, and AstraZeneca was not far behind.

Epidemiologists were in high demand by the media in the year 2020, but vaccinologists were downright celebrities. Sarah Gilbert, the leader of the Oxford vaccine team working with AstraZeneca, even appeared in the August issue of British *Vogue* as one of the magazine's "women shaping 2020." The reserved Gilbert was evidently becoming more comfortable in front of the camera, donning an Armani blouse and Manolo Blahnik boots for her photo shoot.

But one's status as a vaccine developer in the media universe was fraught. It could change at any moment depending on the latest results. In late August, AstraZeneca had shared with Operation Warp Speed early data from elderly adults abroad who had received a dose of the Oxford vaccine — it indicated that the vaccine was only weakly immunogenic in people over seventy. At least four times more neutralizing antibodies could be achieved with a second dose, so a two-dose vaccine looked like the only option. To make matters worse, the manufacturing scale-up at Emergent BioSolutions' factory in Baltimore was preceding more slowly than planned, so there would likely be only fifty million doses available by January 2021. That would now cover just twenty-five million people — if the vaccine was actually approved.

At the time the *Vogue* article was written, Oxford had enrolled eight thousand people in trials it was running for AstraZeneca in South Africa, Brazil, and the United Kingdom, and a readout was imminent. "We can't put an exact timeline on completing the process, but it certainly could be October," Gilbert said.

The U.S. phase 3 results, however, wouldn't be coming until later. For that trial, AstraZeneca was working directly with Operation Warp Speed under the harmonized protocol. The study finally began on Monday, August 31, one month after Moderna and Pfizer started. Operation Warp Speed's leaders knew that there was no way those three vaccines were going to provide the country with three hundred million doses by year's end, and without the Oxford-AstraZeneca vaccine, it might not even crack a hundred million.

During that September 1 meeting between Operation Warp Speed and the FDA, Slaoui had posed a question about AstraZeneca's foreign trials. If the AstraZeneca trial results from abroad were excellent, what would it take for the FDA to accept them? That vaccine was likely to be available in greater quantities than any other candidate, and if the agency wouldn't allow those results, it would make Operation Warp Speed look like Operation Molasses.

Peter Marks didn't know how to address Slaoui's concerns. He remarked that the FDA had had informal discussions with AstraZeneca, but the protocols Oxford used abroad had changed frequently and were "sketchy," he said. The FDA had allowed AstraZeneca's phase 3 study in the United States to proceed based on an analysis Oxford had run on its earlier foreign trials, but the agency had been having trouble obtaining the underlying data from AstraZeneca. "AstraZeneca has not been transparent," Marks said. "We need to know what they've done."

Slaoui felt it was critical to get ahead of the issue and nail down the FDA's criteria for accepting data from foreign trials before the data come out. What would happen, he asked, if Oxford made "an announcement of efficacy that isn't supported?"

Marks sighed. "We have a very challenging situation," he said.

Moderna had pulled together a priority list of twenty clinical trial sites that had the greatest potential to improve its diversity metrics, and the company began a "surge" outreach effort. It revised its recruitment materials to include more images of Black people, altered its phone scripts to help with discussions about the importance of Black participation, and obtained an e-mail database of more than two hundred thousand minority residents who lived near the sites and blasted them with e-mails. Moderna also agreed to increase transparency of its clinical trial, posting its diversity numbers weekly on its website and publishing a patient-friendly version of its clinical trial protocol.

One of Moderna's challenges to building trust in the Black community, as it put it in an internal PowerPoint on September 2, was the "increasing politicization of the trial by media and participants." Along with Surgeon General Jerome Adams, who was Black, Fauci and Collins suggested having weekly calls with the doctors running those clinical trials to encourage them to try harder to recruit minorities. Moderna was enthusiastic about the idea. If the predictions on disease spread were correct, the company still might be able to keep pace with Pfizer and have a readout by October.

Or at least, that was the hoped-for timeline. But a day or two later, Moncef Slaoui burst into Alex Azar's office and said, "You're not going to believe what the FDA did!" He told Azar that Operation Warp Speed had obtained guidance letters that had been sent by the FDA to Moderna and Johnson and Johnson. The FDA told companies that it wanted a median of two months of safety data following the second dose of vaccine before considering an EUA. Sixty days, not forty-two!

To both Azar and Slaoui, it seemed arbitrary and overly conservative amid a pandemic. Why should the FDA hold so much sway on an issue that was hardly cut-and-dried? The military advises the president, but it doesn't decide when the country goes to war. Why should Operation Warp Speed be at the mercy of FDA bureaucrats when it came to deploying the vaccine arsenal?

Azar called FDA commissioner Stephen Hahn and told him he had heard about the letters his agency had sent. "No, we didn't," Hahn said.

"Yes, you did," Azar said.

"No, we didn't," Hahn said again.

"We're literally reading the letter right now," Azar snapped.

Hahn hung up and soon found that, sure enough, Peter Marks and the FDA vaccine leader Marion Gruber had sent those letters to manufacturers without telling him. Marks and his team were drawing their line in the sand, walling off CBER, within the FDA, from the politics surrounding Operation Warp Speed. Hahn undoubtedly knew immediately that he would have leaks to the media and mass resignations if he didn't back CBER's vaccine team now.

In the meantime, vaccine makers had been circling their own wagons. Albert Bourla, the CEO of Pfizer, picked up his phone and called Johnson and Johnson's CEO, Alex Gorsky. Gorsky had been peer mentor to

Bourla when he assumed the leadership role at Pfizer, and the two executives spoke after every new political twist during the pandemic. That week, they discussed the convalescent-plasma saga and the fact that it looked like the FDA was under attack from the Trump administration. This was undermining public trust in the vaccines — *their vaccines.* "We need to go out and say something," Bourla said. Gorsky agreed. Over the next twenty-four hours, they hashed out the text of what they called the "COVID-19 Vaccine Maker Pledge," vowing to follow the FDA's original guidance from June before seeking approval for their vaccines.

Then they began making calls to the other CEOs in Operation Warp Speed. The first person Bourla called was his chief competitor, Stéphane Bancel of Moderna. He asked if Bancel was willing to sign on. Absolutely, Bancel replied. On Tuesday, September 8, the International Federation of Pharmaceutical Manufacturers and Associations published the pledge on its website; it was signed by executives from Moderna, Pfizer, AstraZeneca, Johnson and Johnson, Sanofi, Novavax, and Merck (which had belatedly received Warp Speed funding). The pledge made no mention of the letters the companies had just received, but Bancel and the others had all promised to base their companies' EUA applications on "appropriately designed studies with significant numbers of participants across diverse populations."

Moderna's recruitment numbers were starting to look better — 10 percent of Moderna's new volunteers were Black and 23 percent were Hispanic. Pfizer employed a different strategy to capture more diversity in its trial. Rather than slowing recruitment of whites, as Moderna had, it chose to turbocharge its trial by increasing its size from 30,000 volunteers to 44,000.

But not every company was advancing toward the finish line. AstraZeneca's British trial was halted that week by the United Kingdom's Medicines and Healthcare Products Regulatory Agency. A volunteer in the trial had evidently developed transverse myelitis, an inflammation of the spinal cord that can be caused by viruses and vaccines.

The FDA had been among the last to hear the news. When Peter Marks got a call from *New York Times* reporter Sharon LaFraniere asking about it, he was flabbergasted. The FDA's queries to AstraZeneca didn't receive an immediate response; in fact, the agency first obtained the adverse-event report from colleagues at Japan's Pharmaceutical and Food Safety Bureau. AstraZeneca's position was that it didn't need to disclose the report immediately. "It was very strange," Marks said. The FDA put an immediate stop to the U.S. trial just one week after it had begun.

The hypothetical concerns Moncef Slaoui had expressed to the FDA just over a week earlier about making use of foreign clinical trial data seemed moot. There was a serious safety event to deal with, and Peter Marks and Marion Gruber wanted to take a closer look at all of AstraZeneca's data. Days later, Oxford scientist Sarah Gilbert received the profile treatment in *Politico* magazine. The headline: "Inside Oxford's Vaccine Saga: From Wild Hype to Sobering Reality." For those who already had doubts about the safety of vaccines, the news of the trial hold only added to their worst fears.

22

TRIBULATIONS

On August 15, a Saturday, *Politico* health policy reporter Dan Diamond was driving out to Rock Creek Park in northwest Washington, DC, to hike with a friend. During the drive he checked in with one of his sources on speakerphone. The source made an offhand remark about the pressure that the HHS public affairs chief, Michael Caputo, was putting on career employees at the CDC.

No surprises there, Diamond thought. He knew the CDC's reputation for being prickly — and Caputo's tendency to be a ballbuster. Caputo had once spent nearly an hour haranguing Diamond in a Mob boss sort of way about a story he wasn't happy about. After Diamond's call with his source, he pulled into the trailhead parking lot, scribbled some notes, and enjoyed some much-needed time outdoors.

Diamond wasn't sure there was a solid story there, but he kept digging until he learned something that made his hair stand on end. Caputo's scientific adviser, Paul Alexander, was meddling with the publication of information in the agency's journal, *Morbidity and Mortality Weekly Report*. Known as "the voice of the CDC," the *MMWR*, Diamond knew, was like a "sacred text," meant to provide authoritative advice and information for scientists and health-care workers. If there was one task that the CDC did extraordinarily well, it was producing rigorous documents like this, and it was unheard of for an HHS public affairs official to interfere with it.

Over the next couple of weeks, Diamond obtained e-mails indicating that Alexander was badgering CDC director Robert Redfield to modify two published reports that the adviser claimed exaggerated the risk of coronavirus for children and so would harm Trump's plan to get kids back in school. In another case, Diamond heard that Caputo's team delayed the publication of a study related to prescriptions of hydroxychloroquine, the drug that Trump had touted back in March and April. Alexander wrote that nothing should go out until he had read it and ensured it was "fair and balanced." Caputo, the leak-plugger in chief, didn't yet know it, but he was on the verge of becoming the subject of a major leak.

It was coming at the worst possible time for Caputo. His plans to combat vaccine hesitancy were just getting off the ground. Between May and September of 2020, the share of the population that said it would get a COVID-19 vaccine declined from around 70 percent to 50 percent. A large majority was concerned about the vaccine approval process moving too fast. Caputo was particularly troubled by numbers showing that Trump's own supporters were among the most reluctant to get a vaccine. In addition to securing celebrity endorsements, one of Caputo's more colorful ideas involved contacting the Fraternal Order of Real Bearded Santas, an association of Santa Claus performers, and getting them early access to vaccines as part of an outreach campaign.

In the first week of September, Robert Redfield stopped by Caputo's office after a task force meeting. The CDC director was a bit of a pushover, and he had let Caputo walk all over his agency that summer while remaining friendly with him. Caputo sometimes made Redfield espresso on an electric burner in his office while they discussed their shared faith. Caputo had converted to Catholicism during a dark period in his life. "Why would a benevolent God visit this pandemic on His people?" Caputo wondered out loud. His phone rang and he turned his head to the left to take the call. The corpulent CDC director leaned forward in his chair and poked the right side of Caputo's neck. "Michael, your lymph node is as hard as a rock," Redfield said. "You need to go to a doctor."

The next week, Caputo flew home to East Aurora, New York, for an appointment with his family doctor. He soon got a CT scan. Awaiting results that Friday, September 11, he received a text message from Dan Diamond asking for comment about the way his over-the-top adviser Paul Alexander had been interfering with the *MMWR* publication process.

When Diamond and Caputo spoke a few hours later, it seemed to Diamond that Caputo didn't realize the gravity of Alexander's actions. He defended Alexander, highlighting the fact that he had trained at Oxford University, and explained that the contrarian scholar had simply been "encouraged to share his opinions with other scientists."

Later that evening, Caputo was seething as he read Diamond's story under the headline "Trump Officials Interfered with CDC Reports on COVID-19." Utterly predictable, Caputo thought. He had solved Alex Azar's public relations crisis, and now the media was coming after him instead. But Diamond's story was just the first in a string of negative stories that he knew were coming and that he hoped to kill. The leakers were everywhere, and they were out to get him, out to undermine his president.

The results from Caputo's CT scan arrived in the online portal of his health provider that night, but he failed to check it. The next evening, his doctor called him and said it was critical that he get a biopsy of his lymph node immediately. She couldn't schedule him for the procedure in Buffalo. She suggested he have it done back in Washington.

On Sunday morning, a bedraggled-looking Caputo was sitting on his front porch. He picked up his smartphone and streamed a video of himself on Facebook Live, ripping on CDC scientists. "These people cannot, cannot allow America to get better, nor can they allow America to hear good news. It must all be bad news from now until the election. Frankly, ladies and gentlemen, that's sedition," he said. "I'm not going anywhere. They're not going to run me out. If the president asks me to leave, I will leave. I really want to leave. . . . My health is failing. My mental health is definitely failing. I don't like being alone in Washington. The shadows on the ceiling of my apartment, there alone, those shadows are so long. . . . The problem is in Washington, I can't carry a gun."

Caputo quickly deleted his unhinged post, but not before the *New York Times* obtained it. Just before the *Times* story broke, Caputo alerted the deputy chief of staff at HHS, Judy Stecker, about it. She was infuriated with him, but her tone changed when he told her that he might have cancer. Whatever happened, they would take care of him, she said. That was the paradox of Caputo — he was easy to get mad at, impossible to hate.

After his biopsy at the NIH's Clinical Center, Caputo had one last Zoom call with his staff at the Humphrey Building early on the morning of Tuesday, September 15, a gauze bandage wrapped around his neck. He apologized to them all for making their lives more difficult with his video post.

Not long after, he stepped into a treatment room at the Roswell Park Comprehensive Cancer Center in Buffalo and lay down beneath a hulking linear-beam accelerator shaped like an oversize stapler. The machine produced a dull hum as it blasted his neck with radiation. Caputo had brought his own music to distract him—a recording of the Grateful Dead's May 1977 concert at Cornell University. One of his favorites.

Caputo repeated the process thirty-four times over the next fifty days. He never returned to his job.

Down the hall from the public affairs suite in the Humphrey Building, Alex Azar was facing his own revolt. On September 22, his chief of staff, Brian Harrison, placed the FDA's top vaccine officials on the speakerphone and began the meeting as if expecting an ambush: "This is an informational meeting," he said in his mellow Texas accent. "The secretary is not being asked for his views, and he's not going to give any views."

In order to restore the nation's confidence in vaccines, the FDA had decided to make public the guidance for emergency use authorization that it had privately shared with companies. The previous day, FDA commissioner Stephen Hahn submitted this official guidance document to the White House budget office. Under Trump administration rules, the budget office had claimed the power to approve any new regulations at any agency.* Now Azar needed Hahn and CBER director Peter Marks to tell him exactly what the guidance meant for the vaccine timeline.

"I call it emergency use authorization plus," Marks said. The FDA would require a median of two months' follow-up data from every volunteer in the trial. That meant half of the study subjects would be followed for more than two months and half would be followed for less than two months. Azar did the math and realized that under this plan,

---

* Executive Order 13771, "Reducing Regulation and Controlling Regulatory Costs," was signed by Trump on January 30, 2017.

there was zero chance a vaccine authorization could happen before November 3, Election Day.

"We won't stand on ceremony," Marks said, noting that the guidance wasn't binding. The FDA could lay out its expectations in advance, but it would judge a company's submission for an EUA based on the entirety of the data and the state of the pandemic. Azar knew, though, that the administration had been painted into a corner. Once the guidance was published in the *Federal Register*, the administration would be held to it. If Russ Vought, the budget office director, sat on the guidance or rejected it outright, all hell would break loose among FDA career officials.

Hours after the meeting, details of the guidance were leaked to both the *New York Times* and the *Washington Post*. Azar suspected it was Hahn who had done the leaking in an attempt to force the administration's hand. Trump called Azar to vent. He wanted to fire Hahn. Azar walked Trump away from that path, telling the president that it would interfere with vaccine confidence. "If you fire an FDA commissioner in the middle of this, we'll get a vaccine that no one will take," he said. Then he added, "Peter Marks is in charge of this. You've met him."

"Can you just approve the vaccine?" Trump asked.

"No," Azar said. "It has to work its way through the FDA process, but trust me to work with Peter Marks to get the job done."

The next morning, September 23, Elizabeth Warren, a senator from Massachusetts, shot another arrow into the heart of Operation Warp Speed. Stephen Hahn was testifying in Congress that day, and Warren asked for his thoughts on Moncef Slaoui's financial conflicts of interest. She claimed that Slaoui had failed to disclose shares he held in Lonza, the company manufacturing Moderna's vaccine in New Hampshire. "Dr. Hahn, you and other FDA officials involved in the COVID-19 vaccine must comply with conflict-of-interest laws. You've just told me how seriously you take that. So can you explain to me why Dr. Slaoui should get to play by a different set of rules?"

"Senator Warren, I can't explain the situation," Hahn replied. "I don't have any knowledge of what you described, but I can tell you that we have established a very bright line between Operation Warp Speed and FDA."

"Dr. Slaoui has conflicts of interest. So, to boost the public's confidence, shouldn't he eliminate these conflicts?"

Hahn squirmed in his chair. He was being forced to take part in political theater. One day, he was taking a stand against the White House, and the next he was having to answer for it. That was the plight of the political appointee. "Senator Warren," he said. "I am not aware of the conflicts you're describing, and so I can't comment—I can't—"

"Well, let me put it this way," Warren said. "Congress should strengthen the federal ethics laws to root out this kind of corruption. And the first person to be fired should be Dr. Slaoui."

Slaoui heard about the exchange from Michael Pratt, who was now running the Operation Warp Speed communications team, but it was almost five o'clock before Slaoui watched it inside his office. Pratt had been trying to keep Slaoui out of the limelight as much as possible, approving only a few media interviews while Slaoui's work was under way. That was the entire idea behind Operation Warp Speed, to provide a safe space for the mission outside politics and outside the field of vision of the president. But Slaoui, by remaining silent, had become a punching bag for the Democrats. He was despondent when he called Pratt. "This can't go on. I know I'm not a bad person," Slaoui said. "I want to go on TV tonight!"

Pratt tried to discourage him. He said that there weren't many show options this late in the day. CNBC was about to switch to *Shark Tank*, and Chris Cuomo's show was just terrible. He would also have to get White House approval—

"I don't care if the White House is okay with it!" Slaoui said. Pratt told Slaoui to wait. They would film something that they might be able put out on HHS social media. He grabbed a sign with the Operation Warp Speed logo on it, brought it over to Slaoui's office, and propped it up on the desk behind him. The two men reviewed what Slaoui was going to say and then Pratt pointed his phone at Slaoui, who was clad in a loose-fitting white-and-gray-plaid shirt. The Oppenheimer of Operation Warp Speed had tears in his eyes as he began his impassioned statement. Addressing Senator Warren, he told her that he agreed that Congress was right to demand that conflicts of interests be disclosed but that she was mistaken in her claim that he held shares in Lonza or hadn't been transparent. Although he had briefly been on the board of Lonza in early 2020, he had resigned to take the job with Operation Warp Speed before any shares were transferred to him. Every penny of gain he made from his stock in GlaxoSmithKline would be donated to the government. "I let go of everything I was doing when this pandemic arrived and I was asked whether I do this role to help the American people control this pandemic and save people's lives," he said. "This pandemic is bigger than any one of us. It is bigger than me and bigger than you. Please get the facts right and stop distracting me and the Operation Warp Speed team from doing what we are here to do."

A week later, during the September 30 presidential debate, Trump rejected an assertion that a coronavirus vaccine would not be widely available until the summer of 2021. Americans might still have a vaccine by the election, he said. "I've spoken to Pfizer, I've spoken to all of the people that you have to speak to, Moderna, Johnson and Johnson, and others," he said. "They can go faster than that by a lot." He was banking on a hundred million doses by the end of the year — the most optimistic scenario in Operation Warp Speed's latest confidential forecast.

It had been nine days since Hahn had submitted the vaccine guidance to the White House, and the budget office still had not approved it. Vaccines had become so politicized that New York governor Andrew Cuomo announced that state health officials would hold their own review of the data before distributing the vaccine.

Three days later, on October 2, the seventy-four-year-old president walked unsteadily out of the White House and boarded Marine One. After months of defying warnings and refusing to wear a mask, he had contracted COVID-19. He was having trouble breathing, and his blood oxygen levels had dipped into the eighties. His medical team feared he might need to be put on a ventilator, but they weren't saying any of that publicly. They said they had ordered him to go to Walter Reed as a precautionary measure only. He was installed in the presidential suite on the top floor of the hospital tower, and Michael Callahan got patched in on a call asking for clinical advice. Callahan argued forcefully against giving Trump an interleukin-6 inhibitor, but he was all in favor of famotidine, the generic heartburn medication that had captured his attention in Wuhan. The president would also be put on an all-you-can-infuse buffet that included remdesivir, dexamethasone, and an antibody cocktail.

He recovered over the next few days. But while Trump was busy posing for pictures of himself appearing to do paperwork in his suite, Peter Marks and Stephen Hahn carried out a subversive plan. They had scheduled a meeting of an external FDA advisory panel to discuss the EUA process, a meeting that would be held online for all to see. Although the agency couldn't publish the vaccine guidance as agency policy, there was nothing stopping it from posting the documents as discussion items for the advisory panel.

Azar was angry about this but told Mark Meadows, Trump's chief of staff, it was a fait accompli. "You've got to approve it," Azar said. Hours later, the budget office gave the guidance its stamp of approval. The FDA had just shown the nation that no matter what Trump said or did, the scientists were still in charge, and they were going to decide when the vaccines were ready.

PART 4

ROLLOUT

October 2020–January 2021

23

HOW DO YOU KNOW?

The Mount Calvary Baptist Church sits on a sleepy stretch of road in Rockville, Maryland, across the street from the Polleria Tres Amigos restaurant and the Family Market convenience store. It's an unembellished white-brick building with a peaked roof and the look of a high-school gymnasium. On the evening of Thursday, October 8, Barney Graham stepped inside the large sanctuary and took a seat in a wooden chair next to the altar. Fall flowers of orange and russet stood in vases on two tall pedestals. All the pews were empty. Only the Reverend Barry Moultrie, a big man with a booming voice, and a few other members of the church were present for the videotaping of a special sermon.

The day before, Graham had learned that the drugmaker Eli Lilly had submitted a request for the FDA to authorize its monoclonal antibody bamlanivimab, which had come out of his work with John Mascola at the VRC. The antibody had been identified with the help of his stabilized spike protein. Hundreds of thousands of doses had been purchased in advance by Operation Warp Speed. Bob Kadlec's team at the ASPR would be in charge of getting the drug out to health-care providers in outbreak hot spots. But these kinds of therapeutics were only a stopgap to get the country through the winter. A vaccine was what was needed to end the pandemic. Graham still didn't know if his stabilized spike design, incorporated in most of the Warp Speed vaccines, could do that.

Graham had been chosen as the church's 2020 Men's Day speaker, in part because of his participation every Saturday morning in Mount Calvary's thirty-member men's group. The low-key scientist was, perhaps, the most famous member of the congregation, and it hadn't gone unnoticed within the predominantly Black church that he had mentored Kizzmekia Corbett, a star in her own right. Graham stood up at the podium and removed his mask. Inventing a vaccine was not enough for him. He felt compelled to convince people to take it.

"We are in a once-in-a-hundred-year pandemic," he began. He recited a litany of ills that troubled the world, from a social-justice crisis to the locusts plaguing crops in Africa and India. "It is reminiscent of biblical plagues and suggestive of apocalyptic end-times. It has provoked tensions in political systems on a national and global scale. There is the rise of authoritarianism and nationalism, fueled by fear, that has caused people to become more isolated and uncertain in their thinking and living." He pointed out that the country had lost "iconic warriors of justice" like Elijah Cummings and John Lewis and Ruth Bader Ginsburg. "Many of us are wondering what is going on," he said. "And I have that Marvin Gaye song in my head all the time. How can we know what is true? Who can we trust? And why is all this happening? So I'm going to speak today on the question of: How do you know?"

Graham told the story of Paul the Apostle, a "highly educated scholarly man" who could have tried to persuade people with his eloquent speeches but instead believed that one must come to one's own conclusions spiritually. It was an idea that Graham and his pastor from Nashville, Edwin Sanders, had wrestled with during the HIV days. They called it the faith-versus-fact conundrum; that is, in an environment of mistrust, "providing facts to the mind" does not resolve the questions of the spirit.

Graham spoke of the devastation COVID-19 had wrought on Black and brown communities and reminded the parishioners of the tools that God had given them to come to their own conclusions and how important it was not to succumb to misinformation and conspiracy theories related to the coronavirus vaccine. He gave voice to the horror of the Tuskegee experiment and how it might play into an individual's trust in science or the government. That was something the community understood at "the soul level," he said. "On the other hand, coronavirus is a real threat to your well-being and survival of your families. There may be ways to protect yourself by taking a vaccine and being aware of certain treatment options, which is something you can understand with your mind."

Graham finished up his sermon with a final thought. "I pray that during this time," he said, "we can learn how to listen to each other and connect at a deeper level."

It was now all but certain that Pfizer's mRNA vaccine would be the first submitted to the FDA and potentially the first to be put into arms. Manufacturing the Pfizer vaccine required one hundred and ten days and facilities in three different states. Like Barney Graham, the vaccine was a product of the Midwest. It began inside a plant in the city of Chesterfield, Missouri, about twenty minutes west of downtown Saint Louis.

The Chesterfield Valley used to be called the "Gumbo Flats" because the silty soil turned to stew when the Missouri River overtopped its banks. Those fertile floodplains had been overtaken by tract homes, shopping malls, and Pfizer's five-acre factory and laboratory space that was working twenty-four hours a day to make a vaccine that still had not been proven to be effective.

Inside the Chesterfield plant, a worker began to make a batch of vaccine doses by pulling from the freezer a circular strand of DNA, known as a plasmid, containing the vaccine sequence. The plasmid was then taken up by *E. coli* bacteria, which were cultured in three-hundred-liter plastic bags. Inside each of these bags, enough copies of the plasmid were produced to serve as the raw material for five hundred thousand doses. Next, the bacteria were doused in a chemical that broke open their cells, releasing the plasmids. The plasmids were tested for purity and to ensure that the DNA had been copied accurately. Then the plasmids received a single snip with molecular scissors so that they could stretch out for the next step. This whole process took about ten days.

Next, the linear plasmids were purified and frozen and sent to another Pfizer facility, this one in Andover, Massachusetts. There, they were bathed in enzymes and nucleotides, the building blocks of RNA, which lined up along the length of the linear plasmid. Within hours, the DNA plasmid was transcribed into RNA. The RNA was then separated from the DNA, and the RNA sequence was checked for accuracy.

Finally, the RNA solution was frozen to negative twenty degrees Fahrenheit and flown to Kalamazoo, Michigan, where it was thawed and blended with a combination of four proprietary fats inside a device about the size of a watch face that company employees called the "tea stirrer." This device, formally known as an impingement jet mixer, forced the fats and the mRNA together under four hundred pounds of pressure. The company had been struggling to create larger tea stirrers to scale up production but had abandoned that idea for the time being. Instead, the company built a rack of sixteen computer-controlled pumps to move the ingredients through the process. This "final drug product" was then placed into vials large enough for at least five doses and stored in ultracold freezers that could maintain a temperature of negative eighty degrees Fahrenheit. These vials had to sit for four weeks before a sterility test was conducted.

By the time Barney Graham gave his sermon, Pfizer had its first 1.5 million doses. But the company's scientists and executives were starting to sweat the very real possibility that its vaccine might not come out the big winner they had bet a billion dollars on. The company's trial was still blinded, so Bill Gruber, Barney Graham's old roommate and now Pfizer's clinical leader, couldn't know whether the vaccine was preventing COVID-19 infections any better than a placebo.

Like Moderna, Pfizer had filed its clinical trial protocol with the FDA months earlier. Normally, companies keep details of their protocols secret until a trial is complete to give them a competitive edge, but public pressure led the vaccine makers in Warp Speed to release the protocols in full in mid-September. Now that Gruber and his team had gotten a look under Moderna's hood, so to speak, they could suss out exactly when and how the company planned to analyze its data. As we learned earlier, one major difference between the two companies protocols, which Pfizer had actually known for a while, was that Moderna was going to evaluate its vaccine based on COVID-19 cases that occurred two weeks or more post-vaccination. Pfizer planned to start counting cases a full week earlier. Gruber knew that, in some people, the antibody response would not be fully developed by then, which meant they were more likely to develop infections. That could end up making the vaccine look worse in the final analysis.

As caseloads were starting to rise in the United States and at the company's clinical trial site in Buenos Aires, time was running out for Pfizer to modify its protocol. A blinded trial was like the game show *Let's Make a Deal,* where a contestant can risk his or her winnings for one of the unknown items concealed behind three curtains onstage. Depending on which curtain the contestant chooses, the prize might be a brand-new car or a zonk—a gag item like a live llama or a pair of loafers made out of meat. The last thing a drug company wants to do is get zonked. You can't change your decision after the curtain has been raised, but if Pfizer made a different choice now, Gruber knew, it would be less likely to get zonked. And with the FDA asking companies to wait until they had two months' worth of safety data post-vaccination, it really didn't make that much of a difference.

On Sunday, October 11, just as the FDA's Peter Marks stepped out of the Athenaeum Gallery in Alexandria, Virginia, holding the taut leash of his 110-pound dog Eddie, he got a call from Albert Bourla. The Pfizer CEO told Marks that the company wanted to change its protocol to match Moderna's so that only cases that occurred two weeks or more after vaccination would be counted. Marks was astounded. It was a bold move to change a primary-outcome measurement at this late stage, and it could look like the company had peeked behind the curtain. Not gonna inspire confidence. Over the next couple of weeks, Pfizer and the FDA negotiated a compromise: the company would have to analyze the results at both one week and two weeks post-vaccination.

Before it potentially ended up with a zonk, Pfizer also wanted to seal a deal with Operation Warp Speed for another hundred million doses. But it wasn't clear to the Warp Speed leadership if the company could actually deliver all its promised doses in a timely manner. Part of the problem was that Pfizer's orders to suppliers for parts and materials were still not being prioritized under the Defense Production Act, unlike the orders of the other candidates in the portfolio. Companies would, in theory, have to fulfill orders for Moderna and Novavax before they helped Pfizer. Between the lack of full transparency regarding delivery dates on Pfizer's part and the lack of manufacturing support on Warp Speed's part, it had turned into a lose-lose situation.

Earlier that week, in fact, the Operation Warp Speed team had had a lengthy discussion about Pfizer's manufacturing program. Due to supply-chain issues, the number of doses Pfizer predicted it would have at the end of the year declined from the hundred million doses that Bourla had been hoping for to just fifty million. Moncef Slaoui said that he would call Bourla and make an appeal for more information. "Pfizer needs to be transparent," Slaoui told his team. "We can't help them if they aren't." As Barney Graham said at Mount Calvary, perhaps it was time to connect at a deeper level.

Inside the Humphrey Building, Bob Kadlec pulled General Perna aside. For several months, he had been advocating for a deal to have Merck manufacture the Johnson and Johnson single-shot vaccine. He had been watching the contract manufacturer Emergent BioSolutions struggle with the scale-up process; it had recently had to throw out several batches of vaccine. Merck was an established player with available manufacturing capacity, but it had a cautious company culture. Its two live-attenuated coronavirus vaccine candidates were still in phase 1 trials, and its agreement with Warp Speed had only been finalized in September. "There's Warp Speed time, and there's Merck time," Moncef Slaoui would say at meetings.

Kadlec saw an opportunity to act. "Let's DPA their ass," he told Perna, a half-joking reference to Title 7 of the Defense Production Act, which can force companies to work together and grants them relief from the antitrust act. Perna, however, felt that it went beyond the original mission of Warp Speed, since the government had already signed contracts for eight hundred million doses, and a deal with Merck wasn't going to get the country to the finish line any faster. Merck's factory, like Emergent's, would require retrofitting, which likely would not be completed until the fall of 2021.

Kadlec told General Perna, "My role is to look at the next hilltop and say, 'What do we need to do?'" He called it FUOPS, the army term for "future operations." In spite of everything, he was still the assistant secretary for preparedness and response, after all. It was, he said, within his "scope and powers" to push this effort forward outside of the Warp Speed structure. Alex Azar was supportive.

Kadlec had a number of projects that, as the ASPR, he was quietly working on. For the past six months, he had basically been operating sub rosa. While the administration had effectively stolen his staff and shunted him aside, his loyalists inside the Humphrey Building and Operation Warp Speed continued to report to him and seek out his guidance and assistance. But Dr. Bob's own pragmatic calculations around remaining in the administration had grown more complicated.

He had gotten a bad feeling in the pit of his stomach with the arrival of a neuroradiologist and crank named Scott Atlas, a fellow at the conservative Hoover Institution. Jared Kushner had welcomed Atlas into his own coronavirus meetings in the White House over the summer, and Atlas was becoming a voice inside the official Coronavirus Task Force as well. Atlas was opposed to lockdown measures and widespread testing. "When you isolate everyone, including all the healthy people, you're prolonging the problem because you're preventing population immunity," Atlas told Fox News in July. Atlas believed that the population could cross the threshold of "herd immunity" if just 20 percent of people were infected rather than the 70 percent epidemiologists estimated. On October 17, he tweeted a message to his followers: "Masks work? NO." He linked to an article published by a free-market think tank claiming that mask mandates went against "the principles on which the United States of America was founded."

For several months, Kadlec had been speaking regularly with two old friends, Steven Cash and Kenneth Wainstein. Cash was a former CIA officer and lawyer who once had Kadlec on the roster of his consulting company. Wainstein had been George W. Bush's Homeland Security adviser. Both men were horrified by the damage the Trump administration had done to the country and its institutions. Cash was part of a group of national security experts who called themselves the Steady State and who offered their colleagues in the administration a safe place to discuss the dilemmas they were wrestling with and encouraged them to consider their choices.

In an open letter that summer, the Steady State wrote, "Many of you still serving are facing what is likely among the hardest decisions of your lives: Should you stay in government, hoping to temper the tornado of incompetence and viciousness that seems to be at the core of policy as practiced? . . . The abstract thought boils down to: When do I resign? When do I speak out? When do I become a whistleblower? When do I say 'no'?"

In the upcoming election, Cash and Wainstein, both Republicans, were throwing their support behind Trump's opponent, Joe Biden, and they were lobbying Kadlec to resign from the administration. "You're not one of them," they told him during several phone calls and an in-person meeting.

Kadlec tried to defend himself. "I'm doing what needs to be done for the American people," he said. But Cash and Wainstein were persuasive. They told him that he had the potential to change the course of the election. He could resign during the week of the vote, publish an open letter or an op-ed, and give an unvarnished interview on *60 Minutes*, as Rick Bright had done.

But Kadlec had to wonder if it would help. How many officials like himself had already resigned, and how much impact had those resignations had? He wasn't a beloved household name like Tony Fauci. He wasn't a charmer like Rick Bright. He couldn't paint the world as black and white. The moral and strategic decisions he had made in the fog of war were difficult to convey to the American public. As far as most people were concerned, he would be seen as a crass opportunist, jumping ship at the last possible moment. And there was no part of him that wanted to jump ship. He called Wainstein and reiterated his decision: He was sticking it out. "I have a job to do," he said.

Kadlec had always been a decisive man, because that's what you're trained to be in Special Operations. But he had another side too, one that had gotten the better of him during the pandemic. He was a ruminator. He couldn't stop replaying the past year in his head. The failures of the Trump administration would always overshadow his contributions. *Maybe I* should *resign,* he thought.

Two days before the presidential election, on Sunday, November 1, he called Tony Fauci. Staring out his kitchen window, Kadlec told his old comrade that he was at a breaking point. "Don't do it, Bob. We need you," Fauci said. "I've got your back if you've got mine."

Kadlec hung up the phone and took a deep breath. He set his ego aside. The vaccines were coming any day now, and Dr. Bob still had a mission to support.

24

WINTER WAVE

The day after the election, Wednesday, November 4, it still wasn't clear who America's next president was going to be. Mail-in ballots were still being counted and neither candidate could lay claim to the 270 Electoral College votes needed to win. At three p.m., General Gus Perna provided his weekly video briefing to the Department of Defense leadership back at the Pentagon: Secretary Mark Esper, Deputy Secretary David Norquist, and Mark Milley, chairman of the Joint Chiefs of Staff.

Perna had spent about twelve billion dollars on Operation Warp Speed and expected that number to double by the time the program concluded. Four vaccines were moving forward at a very good pace, he told his superiors. Pfizer and Moderna were approaching submission for an emergency use authorization, and he would be ready to distribute those vaccines within twenty-four to forty-eight hours after approval. UPS and FedEx had built freezer farms at their distribution centers to store Pfizer's vaccines at the required ultracold temperatures, while Moderna's vaccine had been found to tolerate refrigerator temperatures. The vaccine kitting — some two hundred million syringes, needles, and alcohol wipes — would be independently deployed from Tobyhanna Army Depot in Pennsylvania in coordination with each dose of vaccine. The other two leading candidates, AstraZeneca and Johnson and Johnson, could be ready to go in early 2021.

Although the military had been first in line for new vaccines during the 2009 swine flu pandemic, members of the DoD would receive this new vaccine in a phased rollout like the country at large. Across the DoD, there would be nearly three hundred thousand cases of COVID-19 during the pandemic year, but because of the relatively young age of service members, only 1 to 2 percent of those infected ended up in the hospital. As for the broader population, Perna said, almost all the states had submitted their vaccine distribution plans, generally noting that they would await the CDC's recommendations in the coming weeks in order to decide which vulnerable members of their communities should receive the vaccine first.

Perna told Esper this would potentially have a ripple effect on Warp Speed's operational plans. He didn't know how many vaccines he was supposed to ship to each state. Should all states receive doses based on their populations? Or should states with large populations of senior citizens — Florida, for instance — receive more doses? The grunt work of moving vaccines around the country was butting up against these still-unanswered public health and policy questions. "Stay out of that," Esper warned him. "We're not going to have the military be involved in prioritization."

Over the next two days, the uncertainty around the election results dragged on. Joe Biden racked up over 290 Electoral College votes to Donald Trump's 214, but Trump had yet to concede. He claimed baselessly that officials overseeing the count in Pennsylvania were "all part of a corrupt Democrat machine" and leveled charges of voter fraud at states around the country.

Inside Secretary Azar's conference room on the morning of November 6, Bob Kadlec could sense tension among Operation Warp Speed's leaders and scientific advisers. They were there for a discussion about the rollout. Deborah Birx had come from the White House, while Francis Collins and Tony Fauci called in from Bethesda. General Perna espoused the straightforward logic of shipping doses to every state on a pro rata basis. CDC director Robert Redfield agreed with Perna. "K.I.S.S.," he said. "'Keep it simple, stupid.'" The CDC had vetted most of the states' plans for distribution and given them the green light.

Collins brought up the recommendations of the National Academy

of Medicine report he had requested over the summer. It emphasized focusing on the highest-risk health-care workers and then the most vulnerable members of the population.

Birx wondered how vulnerables should be classified. "What role do
comorbidities play?" she asked. "How are we defining *elderly*?" Birx
shared a home with her ninety-year-old parents, and she said she was
watching the coming winter peak with increasing alarm. She had just
come back from a COVID-19 outbreak in North Dakota and feared for
the country. She felt that people in their seventies and eighties, whose
risk of death from COVID-19 was sixty times that of people in their thirties, should be prioritized.

Birx was feeling increasingly frustrated that she hadn't been given
another opportunity to weigh in on the CDC's administration plans
since that meeting in July. In fact, she had been so concerned about the
elderly that she had recently asked Fauci to encourage Moderna's Stéphane Bancel to file a compassionate use application with the FDA that
would get the vaccine to nursing-home residents before there was any
proof that it worked in people. Moderna had declined. A similar request
to Pfizer was also turned down. She chalked it up to the fact that no company wanted to be seen as doing Trump's bidding prior to the election.

Birx was aware of Warp Speed's partnership with pharmacies to vaccinate the elderly in nursing homes, but what about those in senior-liv-
ing communities or multigenerational homes like her own? The government had access to data from Medicare showing where the nation's
fifty-five million people over sixty-five were living. Could the allocations
be targeted to the places and people that needed it most?

The problem, Perna explained, was that it didn't make sense to or-
chestrate a targeting strategy at the federal level. This would require
changing the geographic allocations week by week as new data came in,
but there would be a lag time of one to two weeks before a change per-
colated through the distribution chain. Vaccine availability and vaccination rates were going to change so fast, it was better not to monkey with
allocations. The meeting seemed to end with an understanding that the
first doses had to go out to the states entirely pro rata. The CDC would
simply advise the states on which members of their population they
should prioritize, and the states would run their own local operations.

The issue, however, was not exactly settled in the mind of Deborah Birx. She felt cut out of the decision-making process and still didn't have a full view of how vaccines were going to make it into arms. A few days later, she raised the topic again at a task force meeting led by Vice President Mike Pence. With the winter peak approaching, she argued that the task force should issue clear guidance telling the states they should vaccinate people over seventy along with health-care workers with a high risk of exposure to the virus. Pence warmed to the idea that his task force would decide both how to allocate the vaccines to states and which populations should get them first.

Over the next week, Jared Kushner called Azar to warn him that the topic of distribution was coming up repeatedly inside the White House. Pence seemed to be making a play for Warp Speed; he wanted Perna's team to come to the White House and brief Birx and the rest of the task force.

Ever since Azar's ouster as chairman of the task force in February, the secretary had seen little to like in Pence's brand of bland authoritarianism, and the last thing he was going to do was give him a piece of this historic initiative or let it get torn to shreds by the meddling factions inside the White House. No, this was Azar's baby, blessed by the president himself, and full control had to remain inside the Humphrey Building.

Kushner was in Azar's camp here. "We can't let them federalize this," he said of the task force. Azar and Kushner agreed that state allocations would remain walled off from prioritization. The CDC was already planning to give states its guidance. Certainly, the elderly would be high up on the list. Kushner had an idea: Bring the fox into the henhouse. Operation Warp Speed's four-star general could deliver Pence a formal briefing on the top floor of the Humphrey Building.

The next morning, Saturday, November 7, Trump's bombastic attorney Rudy Giuliani held a press conference about supposed election fraud at the Four Seasons in Philadelphia. Not the hotel. "Four Seasons Total Landscaping," Trump clarified in a tweet that seemed to embody the entire shambolic nature of his presidency. The next day, another meeting with perhaps far greater import for the American public would take place, this one inside a small Pfizer office in Greenwich, Connecticut.

The data from Pfizer's clinical trial was about to be unblinded. Salads and sandwiches were laid out on a long conference table. Soon, chief scientific officer Mikael Dolsten and other executives arrived for their first in-person meeting in months. They nibbled on their sandwiches, keeping their distance from one another. Cases had accrued so fast that by the time Pfizer was getting a preview of its data, there were ninety-four COVID-19 infections among volunteers in the trial, nearly three times as many as originally planned. This could be either a shiny new car or a zonk, depending on how the cases were divided between the vaccine and the placebo group.

CEO Albert Bourla surveyed the room and asked everyone to put bets on what they thought the efficacy of the vaccine would be. Some suggested it would be as low as 60 percent. Others, including Dolsten, guessed it would be around 80 percent.

The leaders of the vaccine team, Kathrin Jansen and Bill Gruber, were in a private Zoom room getting briefed by the independent data monitoring board. They sat through an excruciating introduction as the board told the researchers that Pfizer's trial had been a "high-quality study."

Back in Greenwich, over an hour had passed and the sandwiches had been eaten, but Bourla still had not heard any news from the vaccine team. He and Dolsten were dying. Finally, Jansen and Gruber phoned in. "Good news," Jansen said. "We made it."

The executives cheered and embraced one another. "We have a fucking successful vaccine," Bourla said. A cork was popped and champagne was poured into plastic cups. But what was the number? Just how successful had it been? Finally, Bourla got that news: more than 90 percent effective.

Dolsten rocked back in his chair and threw his arms up. "Unbelievable!" he shouted, looking around at his colleagues, who were equally dazed by that statistic. The company had played by its own rules and won the race.

That evening, as Tony Fauci was attending a socially distant dinner in a neighbor's backyard, his phone rang. It was Bourla. "Tony? Are you sitting down?" he said.

"Hold on," Fauci said as he hurried into the front yard for privacy.

Bourla gave him the news. More than 90 percent effective, he said, repeating the preliminary estimate provided by the data board. "Oh, my goodness gracious," Fauci exclaimed. When the company completed its final analysis of the 194 COVID-19 cases, the efficacy turned out to be 95 percent. There had been just eight infections among more than eighteen thousand people who had received the vaccine. Despite the last-minute freak-out about the new protocol for counting cases, the change had barely made a difference.* Now, everyone in the placebo group would be offered the vaccine, as had been laid out in the company's trial plan.

"Don't tell anyone about it," Bourla reminded Fauci.

News of the success nevertheless trickled down to John Mascola, who passed it on to Barney Graham that night. Graham was in his home office, and he became very quiet. Vaccines against respiratory viruses were never that effective, he thought. Those efficacy numbers put the COVID-19 vaccine in the highest ranks, up there with the measles vaccine. Graham's son Daniel and his two grandchildren, D.J. and Anaya, had come up to visit from Atlanta. They were at the breakfast-room table with his wife, Cynthia. Graham shuffled in, smiling. "It looks like it works," he murmured.

Almost immediately, he retreated back to his office. He collapsed into his chair and sobbed. The tension he had been holding in his body for eleven months was released with the recognition that the vaccine design revolution he had helped set in motion had succeeded. Pfizer's mRNA vaccine was almost identical to Moderna's. For the vaccinologist from Kansas, a win for one was a win for all.

Graham's family came running in, took hold of whatever parts of his big teddy-bear body they could stretch their arms around, and gave him a group hug. "Okay, Bebop," little D.J. said, using his name for Graham. Graham called Kizzmekia Corbett the next day. She had already heard the news from Francis Collins, but they both shared a moment of exhilaration. Moncef Slaoui got the call from Albert Bourla that morning as well — just minutes before Pfizer's press release went out.

The next Sunday, November 15, the process happened all over again,

---

* If you counted cases among vaccinated individuals seven days after the second dose rather than fourteen days, there was one extra infection.

this time for Moderna. John Mascola was sitting in his study anxiously waiting to hear from the NIH's Emily Erberding. "Getting debrief now with ASF in the room," she texted a little after noon, referring to Tony Fauci. A few minutes later, she sent another text: The vaccine was 95 percent effective. Ninety cases of COVID-19 in the unvaccinated group; five in the vaccinated group. No one who was vaccinated developed severe disease — a valuable piece of data that Larry Corey had argued so forcefully to obtain. The bottom line was that Moderna's vaccine, too, was startlingly efficacious.

Mascola rang Graham, but he wasn't answering. After a few minutes, Graham called him back. He listened intently but expressed little emotion. "I can't believe it worked so well," he said, his scientist hat firmly on his head. It was only later, after he set down the phone, that he recognized the heft of it, and once again he grew emotional. Two vaccines, both mRNA, both with his stabilized spike design, had worked extraordinarily well. The shock of it all turned to exuberance, and he realized that, beyond a shadow of a doubt, this vaccine truly had the potential to end the pandemic.

It couldn't come soon enough.

The word from West Texas was that hospitals in El Paso were on the verge of collapse. The county, with a population of about 840,000, was reporting north of two thousand new cases per day. The winter wave was here. On the night of November 15, a woman named Bonnie Soria Najera lost her uncle to COVID-19, the sixth member of her family to die since the pandemic began. On *Good Morning America* she listed every family member who had died so far and recounted the day, months earlier, that she'd gotten a call from the hospital that her father's heart had stopped. They had been trying to resuscitate him with CPR, and it wasn't working. They wanted her consent to give up. "I mean, how do you make that kind of decision?" Najera told the television program's host. "It's your father."

Not long after, Taylor Sexton, who had reserved a precious seven p.m. slot at the gym near his home in Washington, DC, was running on a treadmill, but his phone wouldn't stop buzzing. Sexton, an epidemiologist from Texas, had been working with Bob Kadlec to deliver drugs to the right patients and the right places at the right time. Over the summer, the drug had been remdesivir. Now they were turning to monoclonal-antibody therapies. Eli Lilly's bamlanivimab received emergency authorization on November 19, and their colleagues in El Paso had been desperate to find out when the first batch would show up. The UPS tracking number Sexton had just sent Nim Kidd, the state's chief emergency manager, was dead, and he was blowing up Sexton's phone again. "Any idea when it will be loaded into the UPS system for tracking?" Kidd wrote.

While vaccines were the main event for Operation Warp Speed, they were still at least a month away from distribution. Because of production problems, both Pfizer and Moderna had cut back on their estimated doses of vaccine for the end of the year. But even the most optimistic forecasts predicted that the country wouldn't reach herd immunity until April or May of 2021, which meant that Kadlec's team needed tools like the monoclonals to keep the coronavirus death toll down.

Kadlec had already convinced Texas governor Greg Abbott in October to let him send John Redd, the ASPR chief medical officer, and his medical teams to the Lone Star State, but negotiations with states was always a fraught process. Governors thought that accepting help from the feds was admitting failure, and that was true for both red and blue states. Governors were all pains in the asses as far as Kadlec was concerned, but he had finally persuaded Abbott to allow the medical-assistance teams to staff an overflow site with fifty beds at the El Paso Convention and Performing Arts Center.

With the newly approved antibody, the question for Sexton was, what was the best way to get the drug infused into these patients? Monoclonals were most useful early in the course of the illness of the most vulnerable patients, before they needed hospitalization. Hospitals had to deal with the sickest patients, and they weren't about to divert precious staff or have a patient take up a hospital bed for the hourlong infusion process. Sexton started by sending the monoclonals to El Paso's convention center.

That night at the gym, Sexton stepped outside and made some calls to the staff at the Humphrey Building. The monoclonals were "on the way out the door," Sexton wrote to Kidd. They would be arriving the next day by 12:30 p.m. "Thx," Kidd wrote back.

On the eighth floor of the Humphrey Building, Mike Pence saw a map of his home state of Indiana appear on the screen, and the vice president was quick to remark on it. It was November 19, and General Perna was delivering his briefing to Pence in room 800. The room, which featured a sliding glass door out to a balcony and a panoramic view of the National Mall, was packed with the leadership from across HHS and Operation Warp Speed, and the briefing was going surprisingly well.

The idea that the vice president's task force would alter the pro rata allocation scheme faded away once Perna described how many moving parts there were in the distribution chain and how there was already a clear process in place for the CDC to make its final prioritization recommendations to states.

CDC director Robert Redfield gave a rundown of the timeline. He explained that once the FDA authorized the first vaccine, the CDC's Advisory Committee on Immunization Practices, a panel of outside doctors and health experts, would convene within forty-eight hours to vote on which groups of individuals should receive the vaccine first. Redfield would take that advice into account in making his official recommendation as the CDC director.

There was a record-scratch moment. *What?*

Both Perna and the states needed to know exactly where the doses would be shipped before the vaccine was authorized. Every day they waited meant more lives lost. Ideally, Perna would lock in vaccine-delivery plans to each state by December 4. Bob Charrow, the general counsel for HHS, brought up the fact that the CDC's advisory committee didn't have the legal authority to make a decision about prioritization anyway. ACIP's official role came down to voting on whether Medicare or private insurers should provide reimbursement for various vaccines.

Azar countered that it was important for ACIP to be part of the process. "It inspires public confidence to get that input," he said. He had his beloved mug sitting on the table in front of him, the one with a quote from Winston Churchill: "Keep Buggering On." Even in the pandemonium of the Trump administration, Azar still had faith in creating defined processes in advance of an action, as he had on the issue of using fetal tissue in federal research.

Jared Kushner jumped in. "Can't ACIP meet before the EUA decision?" he asked. The committee could issue its prioritization recommendations, and then states would be able to finalize their plans before the vaccine rollout.

Pence was nodding his head. "That's a great point, Jared," he said. Redfield noted that the advisory committee was already having a meeting that Monday, November 23. He could ask members to provide some kind of preliminary prioritization then, and they could hold their final vote later.

"That's what we have got to do," Pence said.

By the end of the meeting, it seemed as if the crisis had been averted, but then Deborah Birx fired one last missile at Operation Warp Speed. "When you say *pro rata*, what do you mean by that?" she asked.

Perna said he had calculated the pro rata allocations to states based on the overall population. "The entire population is not going to be eligible for the vaccine," she said. The Moderna and Pfizer vaccines were going to be authorized for adults only. The population of children in different states varied from 10 percent to 20 percent, which would effectively skew allocations away from states with older populations. "At the very least you should allocate based on who is actually eligible for the vaccines," she said.

It was a surprising oversight, and Perna agreed to do that. Now the issue really seemed to be settled. Vice President Pence never asked for another briefing.

25

SENIORS FIRST

The next day, Friday, November 20, two armed agents drove their vehicle to the back of the Food and Drug Administration's campus in Silver Spring, Maryland, to Building 71. The modern brick structure with an atrium housed the Center for Biologics Evaluation and Research, the organization led by Peter Marks. After parking the truck, the two agents retrieved a hard drive from the vehicle and hand-delivered it to room G112, known as the Document Control Center. The team there immediately plugged the hard drive into their server and uploaded its contents: Pfizer's application for an emergency use authorization.

Marks soon got an e-mail alerting him that the files were in the system. The head of vaccine review, Marion Gruber, received the same notification. They both logged into the FDA's network using a secure key and began to take in the daunting amount of data. The entire submission amounted to fifteen gigabytes and included the ninety-two-page briefing document and all the raw data that supported it, down to the medical records of approximately forty-four thousand volunteers. Unlike regulators in Europe or Britain, the U.S. FDA reanalyzes drugmakers' raw data before approving a product. Normally, the process for a vaccine approval takes many months of review and multiple meetings and requests for more data, but for the emergency authorization, Marks and his team were planning to compress that timeline down to three weeks.

While Marks worked through the Thanksgiving holiday, eating a sandwich at his desk, Bob Kadlec and his family shared a meal with his mother-in-law, Elfrieda Vrtis. She had suffered a stroke in 2012 and lived alone in Woodbridge, Virginia, about fifteen minutes away from his own home.

The family had all been tested for COVID-19 prior to Thanksgiving. But two days after the visit, Elfrieda seemed groggy and disoriented. The Kadlecs took her to the hospital, and her COVID test came back positive. She seemed to have kicked the virus after a couple of weeks without any complications. But then her symptoms gradually got worse. She suffered from shortness of breath and fatigue for several stressful months over the winter. It seemed to Kadlec as if everyone now had a loved one who had been touched by COVID. (One of his Wolverines, Carter Mecher, would soon lose his mother to the virus.)

Kadlec, who had sat in on those Warp Speed meetings about vaccine distribution, was aligned with Deborah Birx in his belief that the first doses needed to get to the elderly as soon as possible. The number of new COVID-19 cases in the country reported each day had risen from around forty thousand in September to nearly one hundred and eighty thousand over the holiday. Four thousand people were dying every day, and it was only going to get worse.

The problem that Alex Azar currently faced was that the CDC's independent advisory group, the ACIP, was taking a sharp left turn. The group hadn't cast a vote yet, but during its November 23 meeting, its members had begun to settle on the idea that the nation's twenty-one million health-care workers should be at the front of the vaccine line. With the exception of the three million staffers and residents of nursing homes, the group was leaning away from prioritizing the vaccination of elderly or vulnerable adults, including those in group homes. Instead, it was moving toward giving those early doses to the nation's eighty-seven million essential workers, many of whom were from Black and brown communities.

Over the next week, Azar held two calls with CDC director Robert Redfield and the other top doctors in the administration, including Deborah Birx, Tony Fauci, and Francis Collins. He didn't think that the administration should just accept whatever this group of state health officials and academics decided the country should do with its thirty-billion-dollar investment in vaccines. Azar had some of the world's smartest doctors right here, and they were mostly all on the same pragmatic page with him in terms of getting the shots out to the elderly as COVID caseloads soared. They should be helping the CDC make the call, he said.

By the time the advisory committee met for its vote on December 1, Operation Warp Speed was projecting that it would have just twenty million doses by the end of the year. One of the first questions the committee discussed was whether it should recommend prioritizing only a small subset of those twenty-one million health-care workers, perhaps just those in hospital intensive care units during the first phase of the rollout. Then the states could get the remaining doses to the other population groups they felt needed them most.

Jason Goldman, a primary care doctor in Florida who spoke to the committee on behalf of the American College of Physicians, thought everyone working in health care should get the vaccine first. "An ICU physician, while seeing sicker patients, may have greater access to PPE compared to small offices," he said. "I'm greatly concerned about the ability of the small and private practice to be able to get appropriate vaccination, especially from large health-care systems." He added that the "office staff" should also be considered "frontline workers." The final vote just before five p.m. was in line with those doctors' groups.

Azar was at the Japanese embassy that evening and had to leave his cell phone in his limo for security reasons. When he returned to his vehicle, he saw he had missed a call from Redfield. When he called the CDC director back, Redfield was acting squirrelly. He told Azar that the committee had cast its vote and he was about to announce that the CDC would accept its recommendation that states focus on health-care providers in the first phase of the rollout.

"Bob, you are not to do that," Azar snapped. Both the WHO and the United Kingdom had proposed making older adults eligible at the same time as the highest-risk health workers. "Why are we prioritizing a healthy twenty-five-year-old dental technician over an eighty-year-old person with comorbidities?" He reminded Redfield that the ACIP's recommendations held no legal weight. Redfield had promised to have a discussion about this with him and Deborah Birx and other doctors in the administration. "We now need to decide what we recommend," Azar said. He didn't want any more mixed messages coming from his agency or, down the line, from the administration.

"Well," Redfield said, "I'm going to prepare something in writing for review tomorrow."

"Could you at least focus on the health-care providers that are at highest risk and broaden it to the more elderly and vulnerable to be more inclusive?" Azar said. After Azar got off the phone with Redfield, he called two of his staff members, Paul Mango and Brian Harrison. He told them he sensed something weird going on with Redfield. "Make sure he complies with my instructions and doesn't announce anything," Azar said.

The next morning, December 2, the reason for Redfield's evasiveness became clear: He had already signed a memo adopting the ACIP's recommendations and had refused to admit it to Azar. The recommendations were set to be published the next day in the CDC's "sacred text," the *Morbidity and Mortality Weekly Report*.

Deborah Birx and the other doctors were stunned. When Azar saw the announcement, he called Redfield and angrily asked him what the hell was going on. Redfield told Azar that the CDC's senior leadership, including Nancy Messonnier and principal deputy director Anne Schuchat, had brought the memo to him and told him that the CDC director always accepts the ACIP recommendations. "I was told it was within my authority to make this decision," Redfield said.

Like FDA commissioner Stephen Hahn, Redfield had decided, for better or for worse, to stand by his staff. Azar scoffed at his position. He told Redfield that HHS and the White House would provide their own messaging. The secretary's last hope for unity within his organization had now been shattered.

A little after one p.m. on Tuesday, December 8, Bob Kadlec was shivering on the corner of Seventeenth Street NW and State Place — the security checkpoint for visitors to the Eisenhower Executive Office Building. While the rest of the Operation Warp Speed team was already inside for the hastily organized vaccine summit to celebrate the imminent authorization, the White House had failed to send over security clearance for him. It was just one more petty snub that he had to assume had come down from Jared Kushner.

The previous week, Paul Mango had asked Kadlec to lead a Q and A at the event, but the Kushner team had decided to go with assistant secretary of health Brett Giroir, the diehard Trumper who had a minimal role in the program. After that, Kadlec hadn't even wanted to come to the bullshit event, but Azar had insisted on it. *Screw this,* Kadlec thought. He called his assistant and told her to send the car back over to pick him up. He'd rather be working. "No, Dr. Kadlec," she said. "I'll sort it out."

Finally, after forty minutes outside, Kadlec was admitted and took a seat just before President Trump came onstage to expound on the "miracle" of the success of Operation Warp Speed. He began talking about how the administration had put forth a plan to prioritize the elderly alongside health workers. "The ultimate decision rests with the governors of the various states," he added. "We urge the governors to put America's seniors first."

The White House had invited executives from Pfizer and Moderna, but both companies had begged off, apparently not wanting to sully their brands amid the political tempest. When Peter Marks had agreed to be interviewed at the event, he thought it was going to be just a private discussion with state governors, but it had turned into something bigger, as things always did under Trump. Adam Boehler would be interviewing him.

Marks stepped onto the stage, and Boehler thanked him for the hard work he had done over the past few months. He asked him why he thought Americans should trust these vaccines. Marks told Boehler that the FDA's analyses were so thorough that sometimes they revealed things about vaccines that even the manufacturers were ignorant of. The whole process, Marks said, was rigorous and transparent, and the nation would soon have a verdict.

"What happens next?" Boehler asked. "We all hope for authorization. So assuming that happens, assuming you issue the EUA, what happens next?"

"The vaccine will then get into General Perna's hands and they will get it out there," Marks said.

Boehler reiterated how important Marks's work was. "Washington can be a difficult town," Boehler said, "but you and your team and the group here show that this is not a partisan effort. It's an American effort. And so as an American, I want to really say how much I appreciate all your work and your team's work."

The event had also included Azar in a panel discussion with four governors: Greg Abbott from Texas, John Bel Edwards from Louisiana, Bill Lee from Tennessee, and Ron DeSantis from Florida. Azar lauded them for producing "some of the best vaccine distribution plans," which would be "models for the rest of the country." He asked DeSantis what he was thinking about in terms of prioritization.

DeSantis replied that vaccine would be going to both frontline health workers and the elderly. "We think, based on the numbers at the end of December, we can start getting it out into the broader senior population. And then in January really focus on vaccinating as many elderly people as we can. And so we're really excited about it. We've really leaned into it."

"Just imagine that. Every senior in a nursing home in Florida within the next several weeks getting protection," Azar said. A short time later, he noted, "We have of course advice out of the CDC and the advisory committee on immunization practices, but you all as governors have the freedom to decide how you'd like to prioritize."

26

THE N-HOUR

On Friday, December 11, Peter Marks woke at four a.m. and began working in his basement. The previous evening he'd gotten the news that the FDA advisory board had supported an emergency use authorization for Pfizer's vaccine. Now Marks and his team were game to try to move forward so the vaccine could be authorized by Saturday morning. As he reviewed document after document coming in from his team, Marks did his best to avoid checking the latest news about the election. More than a hundred Republican congresspeople had signed a brief filed in support of an application by the State of Texas to have the Supreme Court throw out millions of ballots in four states that Joe Biden had won. Trump was still not willing to concede.

At around eight a.m., Marks got the message from FDA commissioner Stephen Hahn that the two of them needed to call Mark Meadows, Trump's chief of staff. Marks told Meadows the plan: his team was working as fast as it could, and the approval would be issued by Saturday. "I know it can be faster," Meadows replied. He then proceeded to berate Hahn for holding up bureaucratic approvals. "If it's not approved by the end of the day, you should submit your resignation papers instead," he said.

After Marks put down the phone, he called his team together and told them what had happened. "No one wanted to see the commissioner hung out to dry," Marks recalled. No one had forgotten that, by throwing his support behind the vaccine guidance in September, the commissioner had proven he was willing to step into the line of fire to back the FDA career staff over the administration.

Marks was tasked with the final review of the text of the actual label that would be sent to health-care providers with the vaccine. It included information on how the vaccine should be stored, who should not be vaccinated, and what the possible side effects might be. Normally, the label was sent to officials at the CDC and other health agencies to review and then comment on, but in this case, Marks simply got everyone on the phone to go through the wording line by line in real time.

Unlike a drug that has made it through the standard approval process, an EUA label has to include all the details on distribution and monitoring. Someone noticed that the label language said that Pfizer's vaccines had to be kept inside the company's box until the moment they were used. Packed with dry ice, Pfizer's custom-made 975-dose box could keep the vaccines cold enough for only a few days and there was no way it could fit in a freezer. The vaccine doses were obviously going to have to be unpacked. Marks crossed out that line.

Late in the afternoon, he sent the wording over to Pfizer for one final check. Then he picked up the phone and called CEO Albert Bourla to tell him that the FDA was authorizing the vaccine. "We look forward to seeing it deployed," he said.

In preparation for the N-hour, Perna had focused on getting every step right through his repetitive military-style exercises, including shipping dummy doses around the country. They had even arranged for U.S. Marshals to protect some of the first shipments. But now, inside Operation Warp Speed, this Friday-night approval was seen as the worst possible outcome. The general wasn't going to ship the precious vaccines out the next day only to have them arrive at shuttered clinics on a Sunday.

Besides, the 121st Army–Navy football game was taking place that Saturday at West Point's Michie Stadium. Perna sat in his office in the Humphrey Building following the game while making sure everything in the operation was in order. Army beat Navy 15 to 0, leading to cheers in the office. Moncef Slaoui later showed up at the Humphrey Building with a U.S. Navy jersey on, just to get Perna's goat.

• • •

Vaccine D-day arrived on Monday, December 14. At 9:23 a.m., Sandra Lindsay, a nurse at the Long Island Jewish Medical Center, rolled up the left sleeve of her medical scrubs and received the first dose of the Pfizer vaccine in New York State. Lindsay, a Black woman, said, "I believe in science" on a livestream of the event conducted with Governor Andrew Cuomo. "As a nurse, my practice is guided by science and so I trust that." She encouraged everyone who was watching to get vaccinated.

By noon, fifty-five of the initial one hundred and forty-five vaccine shipments had been received at distribution sites, and across the country, vaccine kickoffs were held with elected officials and health workers heralding the beginning of the end of the coronavirus pandemic. A "drumbeat of continuous execution" was the way General Perna put it. By the end of the week, the vaccine would be administered at 1,522 sites. Anita Patel, the pharmacist at the CDC who helped put together the federal program to vaccinate nursing homes, was said to be awed by the way all those exercises Perna had demanded contributed to what was so far a near-flawless rollout. It wasn't what the public health people were used to, but now they saw the value in it.

In Washington State, where the first case of COVID-19 in the country had been reported eleven months earlier, 3,900 doses arrived at the University of Washington Medical Center. The first shots would be delivered on Tuesday to health-care workers and people in nursing homes. Governor Jay Inslee was expecting a total of 62,400 doses that week, and his state had been allocated another 74,100 doses the following week.

But two days after those first shots, Inslee received some disappointing news: the state's second shipment would be cut by 40 percent. "This is disruptive and frustrating," he wrote on Twitter. "We need accurate, predictable numbers to plan and ensure on-the-ground success." Inslee's tweet triggered further anger from leaders in other states around the country. They too fumed that they were being shortchanged.

That's when the finger-pointing began. Alex Azar went on television and explained that Pfizer was having manufacturing problems and the government lacked full visibility into the company's production. CEO Albert Bourla soon lashed out to deny it. He insisted that his company had delivered everything it had been asked for and that there were no manufacturing problems. Pfizer board member Scott Gottlieb even complained on CNBC that the Trump administration had blown it by passing on an opportunity to secure a hundred million more doses of vaccine in November. Gottlieb breezed over the fact that Pfizer had been keeping its distance from Operation Warp Speed for months and had made an about-face only when it saw it needed the government's help to secure all the supplies necessary to manufacture its vaccine.

While it was easy for state governors to make a stink about their allocations, the military had to keep its mouth shut and remain steadfastly apolitical. But General Perna knew he had told the governors that the allocations were just estimates. He knew that vaccine production was like the flow of traffic on a congested freeway, where the slightest perturbation gets amplified, leading all the cars to pile up. Pfizer was already operating at maximum capacity at multiple facilities.

Yet despite all the knowledge about the vaccine industry General Perna had acquired over the preceding seven months, he had made a blunder. He had missed one critical manufacturing step. He knew that after the vaccines rolled off Pfizer's production line and were frozen, they were held for a month, and then doses from that batch underwent sterility tests. What he didn't know was that the vaccine could not be released until forty-eight hours after the FDA received a certificate of analysis from that test. The first shipment had been sitting around for a while so it was unaffected by the two-day discrepancy. It was the second shipment, with the product coming right out of the pipeline, that created the hiccup. Then a winter storm dropped two feet of snow on the Northeast.

On Saturday, December 19, Perna tried to defuse the situation during a news conference following the emergency authorization of Moderna's vaccine. "I want to take personal responsibility for the miscommunication," he said, seated in front of an American flag in a studio in the Humphrey Building. He promised he would try to communicate more clearly in the future. Operation Warp Speed had delivered its first 2.9 million doses of Pfizer's vaccine to the country, and Moderna was moving its product to distribution centers. It would be heading to 3,285 vaccination sites. He believed twenty million doses could still be distributed by the first week of January. "Each shipment is another few yards gained, but every good player or coach knows you need defense along with offense to end the game," he said. Perna encouraged people to wash their hands, wear masks, and stay socially distant. "While we move the ball down the field with vaccines, we will score, we will get to the end zone. It will just take some time to do so."

It was a quiet afternoon in the Humphrey Building. Christmas was two days away and Bob Kadlec was standing inside the Secretary's Operations Center looking up at the dashboard of Tiberius, Operation Warp Speed's vaccine distribution program, named after James Tiberius Kirk, the captain of the *Enterprise* on the original *Star Trek*. Nearly ten million doses of vaccine had been shipped around the country, but only a million had made it into people's arms in the ten days since the first authorization. No one could figure out exactly why it was still going so slowly; perhaps it was just that their expectations had been too high.

Dr. Bob, focused on his mission, was watching hospitals fill up across Southern California, with the worst hot spot appearing to be in the Imperial Valley, an agricultural region near the Mexican border. He was going to deploy his tactical medical team to the El Centro Regional Medical Center to back up that facility and also try to increase the use of monoclonal antibodies. His chief medical officer, John Redd, was also talking to state officials about using the Medicare database to home in on elderly individuals, who would be optimal targets for both monoclonals and the state's vaccination campaign.

That evening after work, Kadlec and Robert Stephan, a friend from his military days, took a table next to the stone hearth at La Chaumière, an unfussy French restaurant in Georgetown. Kadlec had recently placed a wreath at Arlington Cemetery on the grave of one of their own, lost during a helicopter crash in 1993, and the two men reminisced about their service before their conversation turned to the Biden transition, now a month away.

Before the election, Kadlec had expressed optimism that he might be able to keep his position as the ASPR for another six or eight months. Once Trump refused to concede the election, that possibility became less likely. The Biden transition team had added Rick Bright to its COVID-19 advisory board, and Paul Ostrowski of Operation Warp Speed had first briefed the group in late November. But Kadlec and other political appointees were out of the loop. His e-mails offering to privately brief two other members of the Biden camp—including his possible successor as the ASPR—were not being returned.

"It's not like I'm a partisan hack, but that's how I'm getting treated, right?" Kadlec complained as the waiter refilled his wineglass with a deep red Vacqueyras. "If they think my judgment sucks, let's see how they do," he said.

All he wanted was to be able to share the things he had learned to help make the Washington machine run smoothly. Was it finally the end of the road for Dr. Bob, biosecurity legend for nearly thirty years? The world still didn't know about what could have been with America Strong or how dutifully he had served as both the catalyzer and water boy for Operation Warp Speed. He felt defeated. "All I can tell you is I woke up a sadder but wiser man from this experience," he said. Kadlec had achieved some good in these last few weeks, and he still had time to save more lives before his appointment ran its course. But he also knew that he would be tainted for making that choice. "The Trump stink is going to be pretty bad," he said mournfully.

He took a bite of his Dover sole. Then he got a glint in his eye, a hint of optimism, as he brought up what he called his "final act of defiance." Over the past couple of weeks, he said, he had drafted an opinion essay titled "A Marshall Plan for the 21st Century." The 1948 Marshall Plan was the U.S. government's fifteen-billion-dollar program to rebuild Europe in the aftermath of World War II. Through vaccine diplomacy in the post-Trump era, Kadlec saw a chance for the United States to retake its place as a global leader by sending vaccines overseas.

The idea would piggyback on his push to have Merck manufacture the single-shot Johnson and Johnson vaccine, which would cost only ten dollars a dose. The next morning Kadlec was on the phone with Wolverine Matt Hepburn, the vaccine lead on Operation Warp Speed, to repeat what he'd already told General Perna: that without distracting from the overall goal of Operation Warp Speed, he could "get it done."

"Now's the time," Hepburn replied. "All eyes are going to be on Janssen," he said, the name of Johnson and Johnson's vaccine division.

Hepburn was referring to the fact that the doses from Moderna and Pfizer were still rolling out more slowly than anticipated, and the Astra-

Zeneca-Oxford vaccine was unlikely to gain approval in the United States anytime soon. The company had, as Moncef Slaoui anticipated, over-hyped the results from the Oxford-run clinical trials in Brazil and the United Kingdom. It touted a result showing that it was 90 percent effective, but this was based on an analysis of a subset of subjects who had accidentally received a half dose as their first dose. When participants received two full doses, as they had in the Operation Warp Speed trial in the United States, the efficacy dropped to 62 percent. It was a passing grade, but not by much. Johnson and Johnson would top that with just one dose.

Further setbacks to Operation Warp Speed's vaccine forecasts came after Sanofi and Merck announced disappointing early results with Merck abandoning its program. Novavax's candidate, meanwhile, was moving forward slowly, but its relationship with the FDA and Operation Warp Speed had become rocky. Nevertheless, Merck remained hesitant to proceed with the Johnson and Johnson deal, as it was reserving some of its manufacturing capacity for a COVID-19 drug candidate it had spent $425 million to acquire. Kadlec brushed it off as a "salvage" drug that had the potential to save a small fraction of the most desperate patients but wouldn't have the same resounding impact as a vaccine.

One way to close a Johnson and Johnson deal was to build a bridge through NIH director Francis Collins, who had a good relationship with J and J's chief scientific officer, Paul Stoffels. Collins was still a bit sour with Kadlec about how the Rick Bright episode had gone down. Five months after Bright was assigned to lead the diagnostics effort at NIH, he resigned in another huff and tried to throw Collins under the bus with a mean-spirited addendum to his whistleblowing complaint. Few took that one seriously, considering Collins's standing. Collins was well on his way to keeping his appointment under the Biden administration.

He did agree to play the role of intermediary. On January 4, Francis Collins, Alex Azar, and Bob Kadlec hosted the call between Paul Stoffels and Merck's CEO, Ken Frazier. The call went well and the two companies committed to cooperation — to helping both America and the world, as Frazier put it.

With or without Dr. Bob in the picture, a global vaccination plan was already taking shape. In defiance of President Trump, Congress gave four billion dollars to the World Health Organization's vaccines access

program, called COVAX, while the National Security Council was looking into donating surplus vaccine doses directly to other countries.

Early on the morning of January 5, Kadlec drove northwest out of the
capital on Interstate 270. It was a cool, foggy day. The Maryland suburbs
gave way to rolling hills. About an hour outside the city, past Frederick,
he turned onto a smaller highway that wended up into the mist of Ca-
toctin Mountain Park, where the storied presidential retreat Camp David
is located. Azar had been given permission to bring ten people to Camp
David for a morale-booster, and he selected some of the key players
working on Warp Speed, including Moncef Slaoui and Gary Disbrow, the
BARDA director. General Perna was invited but could only participate
over the phone because his wife, Susan, had come down with COVID-19.

Seated around a large table inside the Laurel Lodge, Paul Mango, the
deputy chief of staff for policy, asked the team what aspects of Operation Warp Speed had gone well and what might have been done differently. There was a lot of talk about how successfully the program was
structured, with its "hands-off" oversight, "undistorted" decision-making, and flat organizational structure. That had allowed Warp Speed to
attack what Slaoui called a BHAG: a big hairy audacious goal. Azar said
it was all about the right people, and he couldn't help but mention how
critical it was that Joe Grogan had resigned from the administration in
April and that the vice president and his chief of staff had been kept away
from Warp Speed.

As for shortcomings, Kadlec bemoaned the inability of Emergent and
the other contract manufacturers to scale up for a challenge like the one
this pandemic had presented. Azar and Bob Charrow, the HHS general
counsel, griped that the FDA had reviewed the EUA nearly as stiffly as it
would have a full license. Perna said that one of the biggest mistakes, in
his view, was that the "end state" should have been "shots in arms." They
had focused too much on the intermediate goal of delivering boxes of
doses to the states. Only thirteen million doses had been administered,
fewer than the twenty million doses Azar had been promising for the end
of the year. Criticism of the pace of the rollout was growing in the media.

"The CDC was a challenge," Slaoui said. Azar seconded that.

"What do we need to do in the next two weeks?" Mango asked.

Azar said Operation Warp Speed needed to focus on those shots in arms and on gaining better data on vaccine administration rates. The team concluded that some states were still too focused on rigidly enforcing the CDC's prioritization scheme and that they needed to push governors to be more flexible.

New York governor Andrew Cuomo had even threatened to make it a crime to give shots to people who had skipped the line, which seemed to be a sad recipe for wasting extra doses at the end of the day. Cuomo was so intent on ensuring vaccine equity that he wasn't delivering the vaccine to those who needed it most, the elderly.

By contrast, the leader in the vaccination race, West Virginia, was already well on its way to completing immunizations of its nursing-home residents by making use of independent pharmacies. Like Florida's and Texas's governors, West Virginia's governor Jim Justice had largely disregarded the CDC's advice about health-care workers, choosing to focus the rollout on only those doctors and pharmacists most critical to the COVID-19 response. The state was planning to open ten vaccination sites and start vaccinating all West Virginians age eighty and older along with essential workers, teachers, and health-care workers over age fifty. Justice called the plan "Operation Save Our Wisdom."

The team at Camp David decided that they needed to activate the federal retail pharmacy program they had put together with major pharmacy chains around the country. They should clear any and all bottlenecks for the hospitals and nursing homes administering the still-limited supply of vaccine. And that brought the Warp Speed team to the question of whether it now made sense to release more of the doses they had been holding in reserve for second shots. If vaccine administration was going slower than expected and manufacturing was picking up, perhaps it made sense to reduce the reserve as well? Yes, it did.

Before the meeting broke up and the group went outside to tour the grounds, Azar brought up the issue of "transition governance" of Warp Speed. On January 20, when Azar stepped aside, he wanted to ensure that a career employee, possibly Gary Disbrow at BARDA, would have the power to make any urgent Defense Production Act decisions, at least until the Biden administration appointed its permanent pick for HHS secretary. Everyone hoped things would go smoothly.

27

THE BRAVE ENDURE

Bob Kadlec was back in the Humphrey Building the next morning with a skeleton crew. It was Wednesday, January 6, and a large crowd of Trump supporters were expected for a rally at the White House to protest the counting of the Electoral College votes in Congress. "Be there, will be wild!" Trump had tweeted.

Kadlec figured the protests would potentially create traffic problems and be disruptive to his staff in general. He had asked most of them to work from home. The Secretary's Operations Center would be staffed at a remote site in northwest Washington, DC. One of Kadlec's employees came running into his office a little before noon: "They're marching to the Capitol!" Vice President Mike Pence was inside Congress at that moment, performing the largely ceremonial duty of announcing each state's winner.

Kadlec went up to the eighth-floor conference room and headed out to that balcony with a clear view of the Capitol. A dozen or so HHS staffers were already there. Kadlec saw Azar, who had come from his own office to see how large the crowd was. Some reports were saying there were tens of thousands people, but it looked like no more than a few thousand to Azar. The two men could see smoke bombs going off by one of the Senate buildings, and then Azar thought he saw a banner being unfurled from the roof of the Capitol.

"How on earth did someone get up there?" Azar asked. Then they saw the television reports. The protesters weren't on the roof. They had breached the entrance. At 2:13 p.m. the vice president was hustled out of the Senate Chamber by security as rioters elsewhere shouted, "Hang Mike Pence!" Azar told his staff to go home immediately. Then he and Kadlec grabbed a police scanner from the empty SOC and took it to the secretary's conference room. It was just the two of them listening to the radio chirping as the Capitol Police made urgent requests for assistance.

Azar shook his head without saying a word. Kadlec had never seen him so sullen. "I can't believe it either," he said, just to fill the air. Azar let out a deep sigh and kept shaking his head. Kadlec excused himself and stepped out of the room to check on his own people. One by one, he found all ten of them around the building and told them to follow the buddy rule, making sure everyone made it home safely.

He returned to the secretary's office and found Azar still sitting there, dazed. "Is there anything we need to be doing?" Azar asked.

"Sir," Kadlec said, "all my people have been accounted for."

"Okay, Bob. Thank you."

Kadlec left the Humphrey Building around 4:30 p.m., pulled out onto Independence Way, and stopped at a light, watching a long stream of Trump supporters walking back from the Capitol. The light changed, and they let him pass.

Back at home, he saw on the news that more protests were planned at state capitals that weekend, which meant, potentially, more violence. His daughters were both preparing for their spring semesters in college, and Kadlec, the worrywart of the family, was desperate to keep them under his wing.

Samantha was heading back to William and Mary, where she rented a room on the top floor of a house. When her parents had helped her move in that fall, Kadlec pointed out the best window for an escape in case of fire. "Wait a second," he had said. "They're all painted shut. I'm going to get a knife."

"Bob, please," Ann said.

"Okay, okay," he said. "If anything happens, you'll have to a throw the chair through the window."

As for Margaret, she was returning to the University of Texas at Austin. After graduation, she hoped to join the Marine Corps Officer Candidates School in Quantico. Dr. Bob was proud of his tough little girl even if their relationship wasn't always easy. Hearing about the potential threat of violence at the Texas state capitol, however, he asked Margaret to stay home another week, until after the inauguration on January 20. One of his former Special Operations buddies had even e-mailed him some QAnon conspiracy-theory nonsense—"Join us," he had written—which put Kadlec further on edge. Before he went to work one day, he highlighted an article from the *Washington Post* about the potential violence and left it at the breakfast table for Margaret.

That evening, she told him she was still leaving the next day. He begged her to wait until the weekend, at least. "Absolutely not," she said. "I'm getting out of here." She didn't want to be stuck in Virginia as she had been during the summer, two thousand miles away from her girlfriend.

"If you're not here, I can't protect you," he said.

Margaret winced at the pain in his voice. "You trained me. I'm your daughter."

"You're right," he said. "You're absolutely right."

"Don't worry, I'm not even going to be in Austin," she said. "I'll be dog-sitting in San Marcos."

"Just . . . call me, okay?" Kadlec said.

"Of course I will, Dad."

Barney Graham decided to join Twitter. His contributions to the vaccine had now been featured in *The New Yorker* (twice) and on *This American Life,* among countless other outlets. But he was feeling that it was finally time to tell his own story in his own way. "This is my first tweet," he wrote, "and I plan to recount the previous year events and thoughts about COVID-19 vaccine development." It was Sunday, January 10, 2021, the one-year anniversary of the WH-Human sequence being posted online.

Jason McLellan was among the first to welcome him. Kizzmekia Corbett also gave him a boost. "Y'all!!!!!!!!! My principal investigator THE Barney S. Graham is on Twitter!!! Follow him for a play-by-play of how the last year unfolded," she wrote. "Lol his profile states 'pandemic preparedness' & 'social justice.' It's truly the combo for me." His daughter Annie chimed in as well. "Ok, after that can you tell the stories of the Ebola and Zika and universal flu vaccines," she wrote, attaching an animated GIF of a little girl smiling. A couple of hundred new followers replied to his introduction. Four thousand hit the Like button. "Your expertise is a gift that deserves to be shared," wrote Hadi Yassine, the former postdoc who had caught the HKU1 coronavirus during his pilgrimage to Mecca so many years earlier. In addition to a pandemic-ending vaccine breakthrough, Graham soon had over fourteen thousand Twitter followers to his name.

He dutifully followed up the next day with a tweet about his vaccine-design process. "Spike ectodomain with 2 proline substitutions, no cleavage site, foldon trimerization domain and epitope tags for purification for structure and ELISA; similar protein with AVI tag for mAb discovery probe; protein without tags for candidate protein vaccine."

His wife, Cynthia, was the first to give him a hard time for his outreach effort. "A rewrite of this is definitely in order. . . ." she replied on Twitter.

"Momma stop," his daughter Annie added. "He's talking to his science community right now. He'll talk to us normals another day."

"I told them to leave me alone," Graham said a few days later. "I don't know what I'm doing on Twitter." It was Friday, January 15, and he was stepping into the elevator at the VRC. He was wearing a mask with a repeating pattern of coronavirus spikes on it and was six days away from getting his first shot of the Moderna vaccine. Tony Fauci and Francis Collins had already received their jabs in December.

Outside Graham's office, he saw Kizzmekia Corbett wandering around with her laptop in one hand and a portable light she had used for a video interview with the Discovery Channel in the other. "She's now more famous than I am," Graham joked. Corbett was indeed a lively presence and a powerful science ambassador, especially compared to Graham, who had a tendency to get lost in his own thoughts. When she stopped by his office a few minutes later, he handed her a gift from a coworker that he'd forgotten to share earlier. It was a Styrofoam sphere painted silver with red spikes on it being stabbed with a syringe—a Christmas ornament.

"It's so cute!" she said.

Corbett would be moving on from the NIH soon to become a professor at Harvard, and as much as she was excited about striking out on her own, she was going to miss this place. But there was also a sense that Corbett's and Graham's work together would never be over.

An urgent matter that week had once again brought the team back together. Every time a virus makes a copy of itself, there is a possibility that a mistake will slip into the sequence by chance alone, like the audio on a cassette tape degrading each time it's copied. But unlike with a cassette tape, these mutations may sometimes improve on the original. The virus may become more successful at infecting new victims and multiplying in their cells. One such coronavirus variant had emerged in South Africa, and the data were starting to suggest that it had an R_0 more than 50 percent higher than other circulating strains. Later dubbed the beta variant, it had several mutations on the spike protein, including one right on the tip of the lockpick, the receptor-binding domain. It was caused by a change in a single letter of RNA, which resulted in the replacement of the amino acid glutamic acid (E) with lysine (K) at position 484 in the spike protein sequence:

GNYNYLYRLFRKSNLKPFERDISTEIYQAGSTPCNGV<u>K</u>GFN

As slight as it was, it just might make the antibodies produced by a prior infection or by the Moderna vaccine less effective. The longer the pandemic went on, the more victims infected, the more likely that was. If vaccines weren't rolled out fast enough, not just in the United States but all over the world, the virus would keep evolving and changing, and humanity would be locked in a never-ending arms race. That was the nightmare scenario.

The race was on. Graham and Corbett were once again working closely with Moderna. The NIH was making new spike proteins and testing them against the antibodies generated from the original vaccine design. Corbett wanted Graham's permission to spend four thousand dollars to courier new samples to Moderna over the weekend for one of those mutant-strain assays. He didn't have any qualms. "I was going to do it anyway," she said, revealing a touch of that boldness that had endeared her to Graham in the first place.

Upstairs on the fourth floor, John Mascola was seated behind his desk, planning his next move. Behind him, out his north-facing window, was his future: A plot of land was being cleared next to the VRC for a long-planned expansion that had finally been funded under the Coronavirus Aid, Relief, and Economic Security Act. Taking on Graham's ideas for prototype pathogen preparedness had become Mascola's mission for the future of the VRC. The agency would be working toward preventing the next outbreak of disease X. At the same time, he and his scientists would continue their efforts to create an HIV vaccine, including one they were working on in partnership with Moderna. That was undoubtedly going to take longer than the coronavirus vaccine, perhaps another decade by his estimation, but he would see that the VRC fulfilled the mission that Tony Fauci had initially laid out for it.

The next day, January 16, Peter Marks was out with his dog Eddie for a walk along the Georgetown Waterfront. Found on the streets of Texas with his ribs showing, Eddie, a dogo Argentino, had put on forty pounds of muscle since then and looked like he could easily overpower his rail-thin owner. A few police boats were stationed on the Potomac in advance of the Biden inauguration. Storefronts were boarded up and surrounded by razor-wire-topped fences. Checkpoints manned by military and police had gone up around the Capitol Building, the White House, and the National Mall.

Marks was nevertheless in high spirits, with most of the stress from November and December behind him. He still became worked up, though, when he thought about how his conception of Operation Warp Speed had been so abruptly cast aside. "When I've taken personality tests, I score extremely high on idealism," he said. Eddie proceeded to foul the grass with a turd the consistency of melted ice cream. Marks pulled a plastic bag from his pocket and bagged up the waste as best he could.

In many ways, the biggest winner of Operation Warp Speed was Stéphane Bancel, who had finally proven to the world that Moderna Therapeutics was not vaporware. He had emerged from the pandemic victorious in a race where the three biggest U.S. players—Merck, Sanofi, and GlaxoSmithKline—had failed. He now had his first product on the market, and though he would always have Pfizer and BioNTech breathing down his neck, the work he had done building a broad portfolio was finally going to pay dividends.

During his talk at the virtual J. P. Morgan Healthcare Conference on the afternoon of January 11, he rhapsodized about the success. "I believe that 2021 is going to be an inflection year in Moderna's history," he said. "Before, we believed that mRNA might work . . . now we know this is possible." He noted that more than eighty new human viruses had been discovered since 1980, yet there were approved vaccines for less than 4 percent of them. "Think about that," he said. "In 2019, the worldwide vaccine revenue was thirty-five billion dollars. How big do you think that market could be five to ten years from now if one was able to develop a lot of infectious disease vaccines?"

Based on the agreements the company had already inked with governments around the world for its COVID-19 vaccine, Bancel said it was on track to bring in at least $11.7 billion in revenue by the end of 2021. With its growing war chest, the company was investing in a major way in eight other vaccines, including one against that diabolical virus that Barney Graham had studied so long ago, RSV. Also in its pipeline were additional vaccines being developed with the NIH for HIV and emerging disease-causing viruses, such as Nipah, which, like the novel coronavirus, was endemic to bats. But that was not all, Bancel said. The company was also working to propel more mRNA therapeutics onto the market, including those against cancers, rare diseases, autoimmune diseases, and heart disease. He told viewers that he could not wait to see what the next ten years was going to look like. "I believe this is just the beginning," he said.

Pfizer, in the coming months, would announce that it was going to expand its own mRNA program. It would be winding down its partnership with BioNTech and bringing on fifty new employees to help the company go it alone with all the manufacturing expertise and capacity it had acquired over the course of the pandemic. "We like working with BioNTech, but we don't need to work with BioNTech," CEO Albert Bourla told the *Wall Street Journal*. As for Katalin Karikó, her modified mRNA invention would be further vindicated when news rolled in that the COVID-19 vaccine from a company called CureVac, created without those pricey modifications, failed in phase 3 clinical trials.

At the time Trump left office, the three remaining vaccines in Operation Warp Speed were still stumbling toward the finish line. In the coming months, Johnson and Johnson's vaccine would receive emergency authorization but the contract manufacturer Emergent BioSolutions would have to throw out fifteen million doses of the vaccine after they were contaminated by the still-unauthorized AstraZeneca-Oxford vaccine, produced in the next suite over. But the Merck/Johnson and Johnson deal that Bob Kadlec had pressed for would finally happen in the first weeks of the Biden administration. Novavax's protein-based vaccine, meanwhile, demonstrated 90 percent efficacy in its U.S. clinical trial, but the company wasn't planning to seek authorization from the FDA until September 2021 at the earliest. While Novavax was coming in at the back of the pack in the United States, it hadn't been rendered obsolete. CEPI had funded its scale-up around the world, and its low-cost vaccine still held the potential of putting a dent in the COVID crisis in poorer countries.

Moncef Slaoui wouldn't be staying on to wind down the country's Manhattan Project for vaccines. His contract would end in February, but General Perna would be asked to stay on through the summer of 2021 to marshal the Countermeasures Acceleration Group — it was no longer called Operation Warp Speed — to its end state.

By this point, it was business as usual for Yong-Zhen Zhang and his laboratory at the Shanghai Public Health Clinical Center. On October 26, 2020, he had traveled to the city of Wuhan for the International Conference on Genomics, where he was given an award for "outstanding data sharing." Zhang stood proudly next to the podium, his hands clasped in front of him, while a young woman described the lives he had saved through his principled actions uploading the first coronavirus genome to Virological.org.

In what appeared to be state propaganda, however, she claimed that the initial delay in sharing the sequence was due, not to her country's gag order, but to the slow processing time of a U.S. genomic depository that Zhang had originally uploaded the sequence to. Left unmentioned was the fact that this depository, called GenBank, is primarily a forum to share sequences meant to appear in scientific publications and that the sequences are made public only after they are fully vetted and the submitter provides the final approval.

It was still unknown whether the novel coronavirus had made the leap on its own from a natural reservoir or if it had been helped along by a laboratory accident of tragic proportions. Whatever the case, it seemed more unlikely than ever that the Chinese would be sharing data that would settle the matter anytime soon.

During his final weeks in the Humphrey Building, Alex Azar continued to be dogged by the barbs in the media over vaccine distribution, and he showed up for one of his morning staff meetings with a poem by nineteenth-century author Charles Mackay on his phone. Over the Christmas holidays, he told his team, he had finally caught up on the latest season of the Netflix series *The Crown*. In one episode, the well-coiffed British prime minister Margaret Thatcher recited these particularly apt lines of verse to Queen Elizabeth II, and he now wanted to read the poem to them. "You have no enemies, you say?/ Alas! My friend, the boast is poor." He continued:

> He who has mingled in the fray
> Of duty, that the brave endure,
> Must have made foes! If you have none,
> Small is the work that you have done.
> You've hit no traitor on the hip,
> You've dashed no cup from perjured lip,
> You've never turned the wrong to right,
> You've been a coward in the fight.

The room listened in silent awe, nodding in agreement with the message Azar was sending: that they had to stay strong. The world would come to recognize the blows they had received and the sacrifices they had made. That's, at least, what they had to believe.

At a larger meeting on the morning of January 13, one that included his department heads, Azar told everyone that in spite of the "travesty" of the Capitol insurrection, their work in the Humphrey Building had not been in vain. He himself was always going to be proud of his success on Operation Warp Speed, not to mention the fact that he had achieved his most ardently held policy goals to make drug pricing more transparent. At one point in the meeting he asked his department heads what they felt they had accomplished. Kadlec responded that, even after serving in two wars and five combat tours, his years at HHS had come at a time of "the greatest challenge and peril" for the nation. "It's been a privilege and honor to serve during this unprecedented time," he said. "I will never forget you all for your honorable, heroic, selfless service."

That week, Kadlec started packing boxes in his office. Several staffers offered to help him carry them down, but Kadlec waved them off. "That's my job," he said. They uncorked a bottle of wine around four p.m. and shared a toast to him. Michael Callahan had said he'd be passing through town that week, but he was MIA, as usual. A rumor was going around that he'd received an injection of the Russian vaccine Sputnik V, but he later said that was "categorically not true." He had gotten Pfizer's vaccine. Everyone was a little weepy-eyed, looking around at all the stuff Kadlec had accumulated over these past three years. Kadlec's former chief of staff, Chris Meekins, had given him a framed print of a penguin being carried aloft by a few balloons as two other penguins looked on. *Bob didn't listen to the haters,* read the caption.

At eight a.m. on January 20, Kadlec performed his last duty as the assistant secretary for preparedness and response. It was blustery out on the Washington Mall. He went over to visit his medical teams camped inside a Smithsonian building and then joined them at the base of the Washington Monument in advance of the inauguration. Helping out during the inauguration is one of the traditional functions the ASPR provides, though the teams usually stay in tents on the Mall itself. Kadlec wished them luck and gave them each challenge coins with the logo of the ASPR on them. On one side of the coin were four small images, including one of the caduceus—the battle of the snakes—and another that depicted a vaccine syringe. Wrapping around the edge of the coin were the missions of the ASPR, including the "National Disaster Healthcare System," "Public Health Security Capacity," and, most important of all, "Strong Leadership."

As they were all standing there, Kadlec looked over to the White

House and watched Marine One take off and fly away with the outgoing president. *Good riddance,* he thought. At 11:59 a.m., he walked to his Volkswagen Jetta and started his commute across the Potomac River to Virginia one last time as a member of the Trump administration. That afternoon, 1.5 million Americans got a shot of a COVID-19 vaccine, bringing the total number of doses given in the country to 22.1 million. Few people knew the supporting role Robert Kadlec had played in the race to immunity. Trump had, the previous day, awarded presidential commendations to more than fifty individuals for their "exceptional efforts" on Operation Warp Speed.

Dr. Bob was not on that list.

AFTERWORD

As of this writing, the number of new COVID cases reported in the United States each day has declined from a peak of over 300,000 in early January 2021 to fewer than 15,000. Deaths and hospitalizations are down to an eighth of where they stood during the winter peak. Without a doubt, a significant chunk of that is due to the fact that more than 160 million Americans have received at least one dose of vaccine, and 130 million are fully vaccinated. Pfizer's vaccine has now been authorized for adolescents and young teenagers, and Pfizer and Moderna have both begun the process of seeking full approval for their vaccines from the FDA, which will keep the vaccines on the market even after the public health emergency ends.

On the afternoon of May 12, I rode my bike to Kaiser Permanente in LA for my second dose of the Moderna vaccine. The CDC had stopped recommending masks outdoors, and I saw more smiles on the street than I had in a year. I locked my bike to a sign on Sunset Boulevard and snaked through a line under a tent outside one of the hospital's buildings. Within ten minutes, a nurse had given me the jab along with the second stamp on my COVID-19 vaccination card.

While the pandemic appears to be waning in the United States and in other rich nations, all is not well in poor countries. Global cases actually increased over the early spring months of 2021, largely due to the crisis in South America and India, where some hospitals ran out of oxygen. The world's largest vaccine manufacturer, the Serum Institute of India, had promised to supply low-income nations with hundreds of millions doses this year, but so far it is hoarding them for its own country's people under pressure from the government. Although the United States has begun donating surplus doses to needy countries, herd immunity is now considered out of reach; new variants will keep emerging, particularly in areas where the outbreak is still uncontrolled, and we'll keep having to chase them. The hope is that, one day, perhaps in a few years, the novel coronavirus will be the old coronavirus and as mild as the common cold even among the elderly and infirm.

After President Joe Biden took office at noon on January 20, 2021, he revamped and reinvigorated the entire COVID-19 response by infusing billions more into the vaccination campaign, putting his doctors and scientists front and center, and directing the United States to rejoin the World Health Organization. His administration is both better run and more fundamentally compassionate than the Trump administration, but, in my interviews over the spring, there was a sense of wistfulness among members of the military team inside the Humphrey Building that had bonded over eight turbulent months to achieve the historic mission that was Operation Warp Speed. It was as if they had come back from the war together and had seen some shit, for sure, but life under the new guy didn't offer the same buzz of action. Operation Warp Speed, stripped of its evocative name, felt rudderless and maddeningly bureaucratic. The operation will always be remembered for succeeding in spite of the politics at the time, and it will remain a testament to the grit and ingenuity of the American people.

Being prepared for the next biological threat will require years of sustained investment in basic biomedical research and public health surveillance technologies, not to mention an agile and effective operational response plan. The next outbreak our nation faces may not be another coronavirus. It may not even be another virus. It could be drug-resistant bacteria or even a deadly fungus like *Candida auris,* which has been on the rise inside hospitals over the past year. Another pandemic is not likely, but neither was this one.

There was some talk among members of the Biden team that ASPR might get wound down and its responsibilities folded into the CDC. It was a failed experiment, ASPR's detractors said. That was a view that Bob Kadlec didn't share. The last I heard from him, he was working on his own after-action report, summing up the COVID-19 response in order to do better next time. Senator Richard Burr called him up in March and asked him if he wanted to come over to Capitol Hill again and advise the Senate health committee. He couldn't say no.

Brendan Borrell
June 17, 2021

ACKNOWLEDGMENTS

I am grateful to all the people who trusted me and let me into their lives during the pandemic year, especially to Robert Kadlec, Michael Callahan, Barney Graham, John Mascola, and Peter Marks. Many of them were overwhelmed with requests from other reporters or with the critical work they needed to do, and yet they were always generous and understanding when I came calling again and again. This book might never have happened if I hadn't reached out to Callahan back in March 2020 and asked him if, perhaps, he was doing anything interesting at the moment. Callahan was initially reluctant to speak to me out of fear of harming his Chinese collaborators and his long and fruitful partnership with them, but he ended up being my path to Kadlec, who gave me an exceptionally privileged view of the inner workings of HHS and Operation Warp Speed. The only thing Kadlec asked was that I tell the story truthfully.

I'd like to thank the communications team at the National Institutes of Health, who were always prompt about responding to my interview requests, and Tony Fauci and Francis Collins, who made themselves available to me on multiple occasions. Pfizer also deserves kudos for making available multiple members of their team during this busy time. At the start of my reporting, I spoke to smart people working on a wide range of vaccine efforts who, for one reason or another, didn't get the space they deserved in this book but who helped inform this process immensely. Eckard Wimmer: I still think about you and your wife listening to a classical music CD every morning during the pandemic. I'd also like to thank everyone else who was generous with their time and their knowledge but asked me not to use their names for various reasons. This book is more balanced and more accurate because of you.

I am grateful to my relentless agent, Susan Canavan, who encouraged me to propose a book on the vaccine race back in May 2020 even though I had no idea how I would pull it off. I also have to give a nod to my friend Peter Andrey Smith, who told me she was right. I want to thank my editor, Deanne Urmy, who gracefully taught me what works and what doesn't in book form and explained what it meant when a chapter was "overpacked," as mine so often were. I'll never forget that final month in the trenches before my deadline, shooting chapters back and forth and having daily conversations about what each one was really trying to say. I learned so much from you. I'm also thankful for the editing advice from Alisa Opar, fact-checking from Joy Crane, and copyediting from Tracy Roe, who put her physician's eye to the prose.

Early in the pandemic, my magazine assignments gave me a frequent excuse to cold-call experts and dive into topics I knew nothing about. I want to thank Dan Engber at *Wired,* who published my first piece on the coronavirus vaccine race and gave me, as always, incisive feedback, and Tim Appenzeller at *Science,* who published my first piece on Callahan's exploits in Wuhan. I also have to thank Peter Gwin at *National Geographic,* who let me write about the cruise-ship evacuations and a little more about Callahan, and Michelle Nijhuis, who allowed me to take a very, very deep dive into vaccine adjuvants for the *Atlantic.*

This book benefited greatly from a battalion of eagle-eyed readers. I wish to offer special thanks to Elie Dolgin, who knows far more about mRNA vaccines than I ever will, and to Apoorva Mandavilli, who schooled me on immunity. Charles Sheehan, Yuval Avnur, Matthew Medeiros, Joseph Oleniczak, and Christopher Solomon have all made this book better one way or another. My parents and sisters also volunteered to serve as test subjects and provided valuable feedback throughout.

My amazing partner on this journey, Jodi Kuntz, fed me when I didn't feel like cooking, went hiking with me when I didn't feel like writing, and rolled her eyes at me when I told her I was on the verge of failure. You made this process so much better, and I'll make it all up to you one day. And, Mazzy, I know we haven't spent as much time outdoors as we normally do, but thanks for being there every day, curled up next to my desk and grabbing a squeaky toy every time I had another Zoom interview.

NOTES

My reporting was based on interviews with more than a hundred individuals. I also made several trips to Washington, DC, in 2020 and 2021 to meet with key sources and visit important locations. Every person who appears in a significant way in this book was offered one or more opportunities to speak to me, and many others were consulted who have not been named. A number of my sources, particularly current and former government officials and members of the Trump White House, spoke to me on deep background, requiring that I not cite them as the source of information.

This book uses some reconstructed dialogue based on first- or secondhand knowledge of conversations. Some of my sources also recorded direct quotes in their written notes, which they shared with me. Others were patched in on telephone conversations of third parties or had those conversations described to them immediately afterward. I did not intentionally combine two separate conversations or meetings, but there were a number of cases where multiple meetings on the same topic were held in quick succession and they were difficult to disentangle.

Whenever possible, I corroborated accounts with the help of public and nonpublic documents, news sources, e-mails, text messages, photographs, calendar entries, audio and video recordings, and contemporaneous notes. When hard records existed, such as official transcripts or tweets, I tended to fix minor grammatical or punctuation errors if they were more distracting than enlightening. When records or memories were in conflict, I used my best judgment in going with the most reliable version of events.

The notes below provide sourcing to certain publicly available secondary materials I quoted or relied heavily on and highlight areas in my reporting that required a further layer of transparency around my writing choices.

Prologue

Details about the arrival of the specimen and the sequencing timeline came in part from Charlie Campbell, "The Chinese Scientist Who Sequenced the First COVID-19 Genome Speaks Out About the Controversies Surrounding His Work," *Time,* August 24, 2020. Clinical information about the Wuhan patient and the sequencing data come from Fan Wu et al., "A New Coronavirus Associated with Human Respiratory Disease in China," *Nature* 579 (February 3, 2020): 265–69. The WH-Human sequence and upload information can be found here: https://virological .org/t/novel-2019-coronavirus-genome/319.

I did not speak to Yong-Zhen Zhang, but details about his character and history came from my September 16, 2020, interview with Edward Holmes along with a video interview with Zhang posted on the website of the Falling Walls Conference: https://falling-walls.com/discover/ videos/breaking-the-wall-to-redefine-the-rna-virosphere/.

The likely presence of raccoon dogs, bamboo rats, and snakes at the Huanan market during that period is inferred from Xiao Xiao et al., "Animal Sales from Wuhan Wet Markets Immediately Prior to the COVID-19 Pandemic," *Scientific Reports* 11 (June 7, 2021). Bamboo rats and snakes have also been reported in other studies, and CNN released "unverified" footage from the Huanan market in December 2019 showing raccoon dogs.

Chapter 1: Sounds Like It Could Be Fun

Stéphane Bancel said in a December 17, 2020, interview with Antonio Regalado of *MIT Technology Review* that NIH and Moderna independently came "to the same design." This is apparently a reference to an explicit decision to leave intact the so-called furin-cleavage site, the spot where the spike protein normally gets split into two pieces by the furin enzyme. Graham could have made the spikes more durable by altering the furin-cleavage sequence, but he felt that cut, like a careful incision in a folded piece of origami, would provide the greater flexibility needed for a better immune response. He doubts that Moderna would have made the same counterintuitive call on its own.

Chapter 2: Courage and Doubt

I drew details about Stéphane Bancel's life from published sources and from my own interviews with Bancel's brother, Christophe Bancel, and Bancel's friend Bernard Thierry. Other details about the company's process, including the quote from Hamilton Bennett, were gleaned from Catherine Elton, "The Untold Story of Moderna's Race for a COVID-19 Vaccine," *Boston*, June 4, 2020.

The fact that 119 billionaires were at Davos comes from Tom Metcalf, "Dalio, Dimon and 117 Other Billionaires Descend on Davos," Bloomberg.com, January 16, 2020.

"The only thing one can be sure of . . ." comes from Derek Lowe, "Leaving Moderna," *In the Pipeline* (blog), *Science Translational Medicine*, October 14, 2015.

The statement that the University of Oxford's Sarah Gilbert hated being in the limelight was inferred from descriptions in Stephanie Baker, "COVID Vaccine Front Runner Is Months Ahead of Her Competition," *Bloomberg Businessweek*, July 15, 2020.

The description of the first confirmed coronavirus case came from Michelle Holshue et al., "First Case of 2019 Novel Coronavirus in the United States," *New England Journal of Medicine* 382 (March 2020): 929–36.

Chapter 3: Not Sanctioned, Not Authorized

Michael Callahan spoke about his time on the ground in Wuhan during a March 25, 2020, event of the Medical Exchange Club of Boston; see

Thoru Pederson, "Coronavirus Conversations, in a Time of Logarithm," *FASEB Journal* 34 (May 4, 2020): 6003–5. He subsequently described his trip to me in detail during a five-hour interview on April 13, 2020, and he reviewed his itinerary and provided the exact dates of his trip during a follow-up call on April 23, 2020.

Chapter 4: Our Bugs and Gas Guy

The descriptions of the night of January 28, 2020, and the subsequent negotiations with the CDC came from interviews with ASPR staff and news reports. Additional information about some of the challenges on the ground can be found in the HHS General Counsel's April 24, 2020, findings on the operation.

Issues related to the CDC's stockpile are described in the 2017 HHS Office of Inspector General report "Readiness of CDC's Strategic National Stockpile Could Be at Risk in Case of a Public Health Emergency."

Kadlec would be criticized for his focus on bioterrorism. Protecting the nation against anthrax, in particular, ate up one-fifth of the ASPR annual budget one year. See, for instance, Stephanie Armour, Alexandra Berzon, and James Grimaldi, "Nation's Top Emergency-Preparedness Agency Focused on Warfare Threats Over Pandemic," *Wall Street Journal*, June 9, 2020. Such criticisms leave out the fact that ASPR is not a basic science agency and that there were few late-stage or commercial countermeasures against pandemic threats. "It's easy to say in retrospect, 'You should have been doing this,'" James LeDuc, the microbiologist who leads the Galveston National Laboratory, told me. "Were this a biological weapons attack, he'd be seen as a genius." For more background on the battle between Alex Azar and Joe Grogan over stem cells, see Elaina Plott and Peter Nicholas, "How a Forgotten White House Team Gained Power in the Trump Era," *Atlantic*, June 27, 2019, and Eliana Johnson and Dan Diamond, "Pushed by Anti-Abortion Groups, HHS Restricts Fetal Tissue Research," *Politico*, June 5, 2019.

Chapter 5: Stabilizing the Spike

In Barney Graham, Morgan Gilman, and Jason McLellan, "Structure-Based Vaccine Antigen Design," *Annual Review of Medicine* 70 (January 2019): 91–104, the authors summarized how techniques initially developed to tackle an HIV vaccine ended up being more fruitful with RSV and the coronaviruses. Interestingly, the S-2P design used in the first generation of coronavirus vaccines is not actually as stable as it could be. McLellan's lab subsequently created a spike with six proline insertions — they call it the HexaPro design — that generates relatively more prefusion spike proteins and can withstand heating and freezing. See also Ching-Lin Hsieh et al., "Structure-Based Design of Prefusion-Stabilized SARS-CoV-2 Spikes," *Science* 369 (September 2020): 1501–5. Human trials of a low-cost protein-based version of that vaccine began in Vietnam and Thailand in April 2021.

Chapter 6: Old School, New School

The description of Gregory Glenn's handoff comes from Karen Weintraub and Elizabeth Weise, "The Sprint to Create a COVID-19 Vaccine Started in January. The Finish Line Awaits," *USA Today*, September 11, 2020. The key paper that led to Novavax's vaccine platform was G. E. Smith, M. D. Summers, and M. J. Fraser, "Production of Human Beta Interferon in Insect Cells Infected with a Baculovirus Expression Vector," *Molecular and Cellular Biology* 3 (December 1983): 2156–65.

Chapter 7: Oceanic Outbreak

The budget negotiations between HHS and the White House budget officer were described in Yasmeen Abutaleb et al., "The U.S. Was Beset by Denial and Dysfunction as the Coronavirus Raged," *Washington Post*, April 4, 2020. Some of the events around the shift from containment to mitigation were the subject of a lengthy *New York Times* article (Eric Lipton et al., "He Could Have Seen What Was Coming: Behind Trump's Failure on the Virus," *New York Times*, April 11, 2020). Both of those articles angered many in the White House, who felt they were Azar's revenge for being sidelined from the task force in late February.

Chapter 8: Very Much Under Control

Remdesivir was not as toxic to the liver as Callahan had feared back in February 2020, but even after the drug's full approval by the FDA in October, Callahan said he had never encountered a single COVID-19 patient who met the prescribing criteria.

The issues around the CDC's test kits were reported widely by a number of outlets at the time. See James Bandler et al., "Inside the Fall of the CDC," *ProPublica,* October 15, 2020, for one of the most detailed accounts of the fiasco.

Rick Bright's talk at BARDA industry day can be found here: https://vimeo.com/364826086/e38fe6de5d.

Chapter 9: The President's Powwow

For more on the Life Care Center of Kirkland, see Katie Engelhart, "What Happened in Room 10?," *California Sunday Magazine,* August 23, 2020.

Chapter 10: This Ain't Over

"This ain't over . . ." comes from Tracey Tully, "What a Family That Lost 5 to the Virus Wants You to Know," *New York Times,* June 30, 2020. My account was drawn from several media sources along with my own interview with Elizabeth Fusco on November 25, 2020. The estimates of coronavirus spread in the U.S. population were made online using Gabriel Goh's Epidemic Calculator (https://gabgoh.github.io/COVID/index.html) and reviewed by Daniel Larremore at the University of Colorado at Boulder.

Chapter 11: A Company Is Born

Moderna's story alone could fill a book, and the acrimony of the early days has been covered extensively by other reporters. I've said almost nothing about, for instance, the lipid nanoparticles Moderna used to deliver the mRNA to cells, which is a whole other patent battle. For more information on that, see Ryan Cross, "Without These Lipid Shells, There Would Be No mRNA Vaccines for COVID-19," *Chemical and Engineering News* 99 (March 6, 2021).

"Stable a few seconds in blood" comes from Denise Silber, "When a Harvard Alum Is the CEO of Moderna Therapeutics, in the Race for the COVID-19 Vaccine," *HAE Invites,* June 24, 2020, https://www.harvardae .org/podcasts/2020/6/22/when-a-harvard-alum-is-the-ceo-of-moderna -therapeutics-in-the-race-for-the-covid-19-vaccine.

Chapter 12: Running on Empty

The story about Frank Gabrin and the text messages described come from Alastair Gee, "America's First ER Doctor to Die on the Frontline of the Coronavirus Battle," *Guardian,* April 9, 2020.

For more on Jared Kushner's role in the pandemic response, see Katherine Eban, "'That's Their Problem': How Jared Kushner Let the Markets Decide America's COVID-19 Fate," *Vanity Fair,* September 17, 2020.

According to e-mails obtained by Democrats on the House Oversight Committee, White House adviser Peter Navarro was the main negotiator of the Philips deal, while Adam Boehler was the person who actually signed off on it. See, for instance, Heidi Pryzbyla, "House Democrats Find Administration Overspent for Ventilators by as Much as $500 Million," NBC News, July 31, 2020.

"WE NEED HELP" comes from a Avital Chizhik-Goldschmidt, Jon Kalish, and Molly Boigon, "Burial Society Organization Suggests New Rituals; Brooklyn Funeral Home Overwhelmed," *Forward*, March 31, 2020, https://forward.com/news/442844/borough-park-funeral-home -coronavirus/.

Chapter 13: Masks for Everyone

A senior staffer on the Senate Health Committee explained to me that the $3.5 billion appropriation was meant to support NIH research "in collaboration with BARDA, FDA, and CDC." Part of the rationale was

that ASPR and BARDA had already received funding in the first coronavirus supplemental, while NIH had not. "It was a surprise to us when it
got diverted at the last minute," the staffer said.

The jury is still out on interleukin-6 inhibitors and COVID-19. No
lives were saved in the two trials that BARDA funded. The NIH included
another IL-6 inhibitor, the Eli Lilly drug baricitinib, in one of its mul-
tidrug trials, and it cut hospital stays a single day when combined with
remdesivir. In the RECOVERY trial in the United Kingdom, an open-
label study where doctors and patients both knew which drug was being
used, there was some evidence that Genentech's tocilizumab provided a
benefit in hospitalized patients.

The information on the April 6 Situation Room meeting comes
from Robert Kadlec's memory, and that story was confirmed to me by
Olivia Troye, a former staffer for Mike Pence. On April 8, *Axios* reported
that the program had been killed; see Caitlin Owens and Jonathan
Swan, "Trump Administration Plan to Provide Millions of Free Face
Masks Fizzled," *Axios,* April 8, 2020, https://www.axios.com/hanes
-face-masks-white-house-a09a360f-4d52-4c08-baee-4a85e1f6562f
.html.

"Trump's now back in charge" comes from Bob Woodward, *Rage*
(New York: Simon and Schuster, 2020).

Chapter 14: The Moonshot

Success has many authors, and the inciting incident that led to the
creation of Operation Warp Speed will always be a matter of some
debate. Certainly, it doesn't take a virologist to know that by late March,
the nation needed a significant, organized effort to develop vaccines.
Around that time a group of government outsiders calling themselves
Scientists to Stop Covid-19 was also trying to make headway in pushing
a Manhattan Project–like effort with the Trump administration; see Rob
Copeland, "The Secret Group of Scientists and Billionaires Pushing a
Manhattan Project for COVID-19," *Wall Street Journal,* April 27, 2020.
The bottom line is that the thing that got Warp Speed off the ground
was the initial chemistry between Bob Kadlec and Peter Marks and the
mutual respect they had for each other.

At the same time MP 2.0 and RADx were coming together, NIH director Francis Collins was developing a public-private partnership, dubbed Accelerating COVID-19 Therapeutic Interventions and Vaccines (ACTIV). There were initially some crossed wires between the overlapping programs and the companies, not to mention a jostling among the players for money and control that I haven't described in all its gory detail. By April 29, Kadlec had branded MP 2.0 the "effector arm of the ACTIV effort," giving the NIH a role as the principal adviser on the clinical trial design. BARDA, meanwhile, became the interface with the DoD, the CDC, and the other agencies on manufacturing and distribution.

Chapter 15: Return to Power

As of June 2021, there has not been a final determination regarding the allegations in Rick Bright's whistleblowing complaint. He is now the senior vice president of Pandemic Prevention and Response at the Rockefeller Foundation.

Chapter 16: Horse Race

The harmonized protocol was first laid out in Larry Corey et al., "Briefing Document on Involvement of NIAID in Development of Public-Private Partnership for the Clinical Development and Evaluation of COVID-19 Vaccine for US and Globally," National Institutes of Health, April 9, 2020.

For more on the debate over the origins of COVID-19, see Katherine Eban, "The Lab-Leak Theory: Inside the Fight to Uncover COVID-19's Origins," *Vanity Fair,* June 3, 2021.

Chapter 17: Battle Rhythm

Carlo de Notaristefani's June 4, 2020, visit and subsequent report were described in Sheryl Gay Stolberg, Sharon LaFraniere, and Chris Hamby, "Top Official Warned that COVID Vaccine Plant Had to Be 'Monitored Closely,'" *New York Times,* April 7, 2021.

Chapter 18: Masters of the Protocol

"Provided no evidence . . ." comes from Kizzmekia S. Corbett et al., "Evaluation of the mRNA-1273 Vaccine Against SARS-CoV-2 in Non-human Primates," *New England Journal of Medicine* 383 (2020): 1544–55.

Chapter 19: Three Hundred Million Doses

The reporting on the relations between Operation Warp Speed and Pfizer comes from multiple confidential sources inside the government and the notes from a participant showing that this frustration continued into October. According to one report (see John Lauerman and Riley Griffin, "Pfizer to Supply U.S. with 100 Million More Vaccine Doses," Bloomberg, December 23, 2020), the company ended up reversing course and requesting the use of the Defense Production Act in mid-December. "If you ask the government to give you the Defense Production Act support, the government has, by definition under law, the ability to reach into whatever you're manufacturing," Moncef Slaoui was quoted as saying. In an interview with me, however, Pfizer CEO Albert Bourla insisted that the company always provided Operation Warp Speed with all of the manufacturing details it needed. "They knew exactly what was happening," he said. "They had very good visibility on everything."

"You ready?" comes from WTOC-TV's broadcast of Dawn Baker's vaccination, which was posted on Facebook on July 27, 2020.

Chapter 20: Plasma Drive

"He said they just had two reports of polio . . ." comes from Edward Shorter's 1987 book *The Health Century,* which also describes the origins of the biologics laboratory that became CBER. Additional details about the Cutter incident can be found in Paul Offit, *The Cutter Incident: How America's First Polio Vaccine Led to the Growing Vaccine Crisis* (New Haven, CT: Yale University Press, 2005).

Chapter 21: Trials

For an excellent summary of AstraZeneca's troubles, see Rebecca Robbins et al., "Blunders Eroded U.S. Confidence in Early Vaccine Front-Runner," *New York Times*, December 9, 2020.

Chapter 22: Tribulations

In addition to the *Morbidity and Mortality Weekly*, Michael Caputo and other members of the administration, including Deborah Birx and Brett Giroir, rewrote certain public guidance documents issued by the CDC. After the Biden administration took office, the new CDC director, Rochelle Walensky, ordered a review of these documents that identified three that had been "developed or finalized outside of the agency" and essentially downplayed the pandemic.

Chapter 23: How Do You Know?

My description of Pfizer's manufacturing process was crafted with the excellent interactive story by Emma Cott, Elliot deBruyn, and Jonathan Corum, "How Pfizer Makes Its COVID-19 Vaccine," *New York Times*, April 28, 2021, and Sue Halpern, "Why COVID-19 Vaccines Aren't Available to Everyone," *New Yorker*, March 13, 2021. I obtained the detail about the tea stirrer from Joe Watson, "Making Moonshot: How Pfizer Makes Its Millions of COVID-19 Vaccine Doses," *Cincinnati Chronicle*, April 2, 2021.

Chapter 27: The Brave Endure

Moncef Slaoui departed Operation Warp Speed on February 10, 2021. That month, his former employer GlaxoSmithKline received a letter from an employee alleging that he had sexually harassed her while he was serving on the company's board. Following an investigation, GSK announced on March 24, 2021, that it had substantiated the allegations, was severing ties with Slaoui, and would rename the Slaoui Center for

Vaccines Research in Rockville, Maryland. In a statement released that afternoon, Slaoui apologized "unreservedly" to the employee. "I will work hard to redeem myself with all those that this situation has impacted," he wrote.

A description of Yong-Zhen Zhang's award and a link to a video of the presentation can be found at EurekaAlert, https://www.eurekalert.org/pub_releases/2020-11/g-cdh111020.php.

INDEX